Ants

BRITISH WILDLIFE
COLLECTION

11

Ants

The ultimate social insects

Richard Jones

BLOOMSBURY WILDLIFE
LONDON • OXFORD • NEW YORK • NEW DELHI • SYDNEY

To my colony or, as it is better known, my family, with love.

BLOOMSBURY WILDLIFE
Bloomsbury Publishing Plc
50 Bedford Square, London, WC1B 3DP, UK
29 Earlsfort Terrace, Dublin 2, Ireland

BLOOMSBURY, BLOOMSBURY WILDLIFE and the Diana logo are trademarks of
Bloomsbury Publishing Plc

First published in the United Kingdom 2022

A catalogue record for this book is available from the British Library

ISBN: HB: 978-1-4729-6486-1; ePDF: 978-1-4729-6489-2; ePub: 978-1-4729-6488-5

2 4 6 8 10 9 7 5 3 1

Page layouts by Susan McIntyre
Jacket artwork by Carry Akroyd

Printed in China by RRD Asia Printing Solutions Limited

To find out more about our authors and books visit www.bloomsbury.com and sign up
for our newsletters

HALF TITLE: A *Myrmica* worker grappling a succulent caterpillar back to the nest.
FRONTISPIECE: *Lasius niger* workers tending an aphid colony.

Contents

Preface

I have always been fascinated by the smaller, less showy insects. I got my interest in natural history from my father, Alfred Jones (1929–2014), an amateur but expert botanist and entomologist. When my family moved from urban terraced Croydon to a sunny detached house in Newhaven, in Sussex, in 1965, the great vistas of the South Downs, Sussex Weald and English Channel were laid out before us, and we leapt at them with extensive family rambles most Saturdays and Sundays from March to October. Every trip out was a voyage of discovery into the marvels of nature. By the time I was 10 years old I was already finding beetles and bugs, flies and fleas, to look at under the hand lens, having eschewed what I considered the rather unnecessarily gaudy butterflies and moths. Ants were there in the mix too.

Before we owned a car those family trips took the form of picnic hikes over the Downs (literally just out the door and up the hill) or short journeys by bus and train to Friston Forest near Eastbourne, Mount Caburn on the outskirts of Lewes, or particularly the famous Victorian naturalists' destination, the Abbot's Wood complex at Polegate. It was here, in the extensive mixed broadleaf and conifer woodland enclosures of Ogg's, Nate, Gate, Abbot's, Folkington and Wilmington Woods that I had my first real close encounter with ants. These weren't the tiny black or yellow insects I knew from the garden, these were wood ants, and they were giants. They also made awe-inspiring heap-nests – as tall as me they seemed – and on a warm spring day they emerged from the depths and gathered together in writhing black knots in the sunshine. I was transfixed by the sight. I was never squeamish, having been on face-to-face and hand-to-mud terms with natural history all my young life, but I can still recall my skin crawling at the sight of the squirming multitude. And as I stood silently watching this marvellous spectacle, I also became aware that I could actually hear the ants moving all around me as their assembled feet clattered across the dead dry leaves in the undergrowth. This was a true power and strength I had not previously credited to tiny, puny insects.

OPPOSITE PAGE:
A wood ant *Formica rufa* stands sentinel on a waxcap fungus.

While checking my copy of Westwood's *Introduction to the Modern Classification of Insects* (1840) to see what he has to say about ants, I get to page 227 and find a winged male black ant, *Lasius niger*, pressed flat in the gutter.

Ants are tiny, but their numbers are almost beyond computation. It is this that makes them such a powerful ecological force in the world. Even though we have a rather impoverished ant fauna in the British Isles (only about 50 species, compared to nearly 200 in France), individual ants are still numerous enough to be important landscape modellers. No scenic view is complete unless it shows the regular polka-dotting of ant hills in a rough grazing meadow. And although some ant species do seem to occur everywhere, under each stone or log, we also have several scarce species, just on the very edges of their European ranges, and these offer deep insights into habitat quality and nature conservation.

Perhaps the deepest fascination with ants comes from an understanding (or if that's too complicated, then at least an initial examination) of their complex nesting behaviours, their foraging strategies, their interactions with each other and with other animals, and with the strange three-gendered colony structure that involves males, females and 'neuter' workers. These peculiarities lift ants above the normal study of entomology. There is some complex science here, enough to fill the time and try the patience of many lifetimes, but rather than complicating, I hope that the few examples I have chosen are illuminating.

Although this book is ostensibly about the ants of the British Isles, I constantly make detours and diversions across the globe to look at examples of the far greater diversity of ants that live beyond our shores. I crave the indulgence of readers, who I hope will make these mental journeys too. For many hundreds of years an accepted fact of ant superstition and folklore – that these small insects collected seed grains – was doubted and debated by an insular northern European intelligentsia who failed to look beyond their own non-harvesting ant faunas.

In entomology there is always so much new to learn. And although there are 'only' about 12,500 ant species in the world (compared to approaching half a million beetle species), ants provide more than enough food for thought. It has been known for me to be sitting quietly in my living room, engrossed in reading some obscure book or scientific paper on insects, and suddenly to come across a new fact or concept that makes me think 'Wow!' I then glance up to find all my family looking at me, and I realise that I have said it out loud. It happened countless times while I was researching this book.

I chose the subtitle partly because *Ants* is such a short title on its own. Also, ants *are* the most social of all the insects. In the famously social bees and wasps, most species are actually solitary, with lone females working individually to each create a small tunnel nest, stock it with foraged food and lay their few eggs in it. They will die off before their offspring take flight the following season. A few, like bumbles and yellow-jackets, are annually social, with overwintering mated queens eventually creating large nests full of workers, but these colonies are short-lived — they wain and vanish after producing new queens and males. A few other wasp and bee groups are perhaps on the first rung of the social ladder, with some subordinate females helping out their matriarchal mother in the temporary nest.

In these very familiar insect groups only the Eurasian honeybees (tribe Apini) and tropical stingless bees (tribe Meliponini) are truly, fully, 'eusocial', creating permanent long-lived, potentially immortal colonies with large numbers of female but non-reproductive workers. By contrast, almost *all* ants are eusocial insects. Despite honeybees being the best-known colony creators, they are actually the oddities; ants should really be regarded as the paragons of insect social behaviour. I offer this book, appropriately subtitled, in support of that proposition.

Richard Jones
East Dulwich
August 2021

What's so special about ants?

Ants are small – so very, very small. Indeed, ants are the epitome of insignificance. They are the default small animals to which we compare migrating wildebeest viewed from a hot-air balloon safari, or the crawling masses of humanity seen from the viewing platform of the Shard or the Eiffel Tower. They are the caricature moving dots in the picnic-raid cartoon strip, the unseen irritant in the underwear. The comedy of the 1960s Hanna-Barbera cartoon animation *Atom Ant* was that, despite his small speck size, he had fantastical superpowers of strength and speed, when in truth ants are small and feeble and can easily be squished. But ants do have a real superpower which makes them among the most important organisms on the planet – what they lack in size and individual strength, they more than make up for in sheer numbers and sophisticated coordinated behaviour.

Consider a rough grazing pasture anywhere in the British lowlands. Some of the roughness comes from the diverse plant architecture that such unimproved fields enjoy, and part comes from the random trampling of the stock animals, but the irregular landscape is dominated by the regularly spaced mounds of ant hills constructed by the yellow meadow ant *Lasius flavus*. These are creatures barely 3–4mm long, and the mounds, some of which would take several wheelbarrows of soil were they to be sculpted by human workers, range from half-football to half Smart Car in size. Some have been present for decades, many for close on a century. Each might contain 10,000 ants, churning the soil, altering the substrate sufficiently that the mounds have their own specialist flora.

As a boy ranging the South Downs behind my parents' East Sussex house, I rather took these ant hills for granted, playing a kind of

OPPOSITE PAGE:
Early in the spring sunshine, wood ants *Formica rufa* leave the depths of the colony and huddle together to warm up in the sunshine.

11

A tale of two cities — multiple domed ant cities in Prior Park, against a distant backdrop of Bath and (right) the tiny but numerous yellow meadow ant *Lasius flavus* that builds them.

hillside hopscotch on them with my school friends. The only time we took care to notice them was in the rather foolhardy sport of down-scarp one-man corrugated-metal toboggan suicide. While sitting on the large flexible galvanised metal sheet, holding each edge up to form a trough-like sledge, the idea was to launch from the top of the grassy slope above Poverty Bottom and glide gracefully (actually hurtle recklessly) down the steep hill. This seldom happened without incident. Hitting the not insubstantial mass of a *Lasius flavus* ant hill head-on had the unfortunate effect of stopping the metal sheet dead, while the pilot was propelled some distance onwards by considerable momentum. Failing to adequately anticipate the collision and let go of the metal sheet's razor-sharp edges at the right moment inevitably meant palms and fingers gashed open as they were dragged along the rusty rims. With hindsight I think we were all lucky that we didn't lose any digits or get serious blood poisoning.

I treated the giant heaving heap-nests of the wood ant *Formica rufa* with far greater respect. At up to 11mm long, this is Britain's biggest ant, and with approaching half a million individuals in a colony, it also builds Britain's largest ant nests, easily a metre high and two across, sometimes larger. Their long trails across the sandy paths and dead leaves (where you can also hear them walking) are a constant fury of activity as they come and go, often carrying the remains of small insects they have killed or scavenged. Dismembered bush-crickets, beetles, caterpillars and butterflies are taken back, trophies, into the thatched ant city. Wood ant size and numbers make them a significant presence in woods, and with estimates of 6 million prey items a year from half a hectare, their impact on local invertebrates cannot be underestimated. Their potency in forest ecology has led to them being protected by law in several central European countries.

And 'flying ant days' are now national news. These mass aerial releases of new males and queens, usually of the ubiquitous black pavement ant *Lasius niger*, fill the air with countless millions of flying insects. The flying ant morsels are manna to the swifts, swallows and martins that scream after them, but a potential breathing or blinking hazard for cyclists, and a real visibility nuisance to motorists running low on screen-wash.

Giant wood ant heap-nest in the Caledonian Forest of Scotland.

A *Lasius niger* queen ready for flight on a warm July afternoon.

How's this for a strange coincidence? On the day that I wrote that paragraph, 4 July 2018, at the very start of putting together this book, it was flying ant day in Nunhead, where I was writing, and East Dulwich, to which I cycled home in the afternoon. I didn't swallow any ants on my way, but plenty bashed into my face and got tangled in my eyebrows. It's a gentle traffic-free cycle ride through Peckham Rye Park, so there wasn't too much chance of me being dangerously distracted in front of a speeding juggernaut. However, later that same day the number 2 women's tennis seed Caroline Wozniacki lost to relative outsider Ekaterina Makarova; magnanimously Wozniacki did not blame the flying ants for her surprise defeat, but earlier she had asked the umpire if anything could be done about the 'bugs', and the media coverage was full of images of her swatting at the hordes of winged ants sweeping across Court Number 1 at Wimbledon. There was another media circus event in July 2019 when clouds of flying ants confused the meteorological office by appearing on the increasingly sensitive radar as 'rain clouds' over Brighton, Bournemouth and Winchester.

Some ant world records to set the scene

Leaving British ants for a moment, it is worth examining some of the quite frankly mind-boggling numbers that are often bandied about by entomologists from around the world, where ant research is more widely funded as an adjunct to nature conservation or as a direct response to agricultural losses caused by these ubiquitous creatures. For example: roughly one-third of the entire animal biomass in the Amazon rainforest is reckoned to be ants and termites, each hectare home to 8 million ants and 1 million termites. Together with bees and wasps, these make up more than 75% of the total insect biomass there (Fittkau & Klinge 1973). Ant activity in the ground is reported to move as much soil as do those traditional subterranean labourers the earthworms – 7.4 tonnes per hectare per year in the North American *Formica cinerea*, equivalent to moving 15cm of subsoil to the surface in 106 years (Baxter & Hole 1966). In another study 5,000 Floridian harvester ants *Pogonomyrmex badius*, each only 3mm long, moved 20kg of sand in five days when they excavated 10m of tunnels (Tschinkel 2004). In tropical rainforests less than 0.1% of nutrients are deeper than 5cm below the soil layer,

Slightly fanciful depiction of 'ant nurseries', from a Victorian children's natural history book (Tucker 1889).

but *Atta* leafcutters transport significant quantities down to nearly 6m. Where leafcutters are active, transport of key nutritional elements from decaying litter to tree roots can be increased nearly 100-fold (Haines 1978). Ants are busy the world over, digging, earth-moving, cycling and recycling nutrients.

A large subterranean leafcutter (*Atta/Acromyrmex*) nest in Central America can contain upwards of 6 million individuals living and working together, and consists of multiple chambers and cavities that would fill a camper van, or two. When I came across a large clearing on the floor of the dense jungles of Carara National Park in Costa Rica in August 1991, it took me a few minutes to work out what I was looking at. Under the tall trees the bare dark soil was devoid of all other plant growth across an area as big as a tennis court; every metre-and-a-half or so there was a small volcano-shaped mound 10–20cm high, and at each peak was a narrow round central hole disappearing deep into the ground. Eventually I discovered that one of the volcanoes was active: near the edge of the clearing a single hole was receiving a constant stream of large leafcutter ants, each bearing neatly clipped green leaf portions down into the darkness. At a rate of about one ant every couple of seconds, 5–10cm apart, they descended endlessly, a robotic procession, into the gloom. Despite half an hour of searching I never discovered whether there was another mound where a burden-free stream of eager ants was exiting onto the forage trail. For the time being this seemed a one-way journey. Occasionally, from one of the other volcanoes, a lone ant would emerge with some dead brown plant remains in its jaws; it dumped this detritus over the rim of the mound, then vanished back into the hole. My mind still reels to consider the vast insect metropolis that had been beneath my feet that day, or the massed labour it had taken to clear the forest floor thereabouts.

What makes the awesome wonder of a leafcutter ant nest even greater is the fact that each ant, though working individually as a discrete living organism on its own, is also part of the complex machinery of the community. Every ant is a well-oiled cog in the mechanism. But the cogs come in different shapes and sizes – even though they are all genetically virtually identical, a nutritionally and hormonally controlled caste system of tiny nurses, large harvesters and even larger soldiers divides the labours of the colony according to their individual talents. Every ant has its place, and its work cut

A South American leafcutter *Atta* species takes its fresh cargo home to the nest.

out for it. The complicated and (more importantly) coordinated behaviours of exploring, foraging, nest-building, nest defence, brood care, fungus farming, food sharing, air-conditioning and colony hygiene are not carried out under anything approaching individual sentient thought, but are nevertheless communicated through the colony by long-distance chemical trails, local hormonal osmosis and one-on-one tactile signals. A sort of collective sociochemical subliminal telepathy unites the many disparate ant units to one overpowering future objective – colony reproduction. The leafcutter ant nest is, truly, a wonder of nature – and throughout the literature metaphors abound as ant colonies are likened to armies, cities, nations, civilisations.

If 6 million Central American *Atta* leafcutters make an ant city, then 306 million specimens of *Formica yessensis* in 45,000 interconnected nests through 2.7 square kilometres of the Ishikari coast of Japan's northern island Hokkaido (Higashi & Yamauchi 1979) make an ant conurbation. A single column of driver ants (*Dorylus* species) in tropical east Africa can have 20 million individuals scattered across several loosely connected bivouacs through several hundred metres of jungle – a diaspora perhaps. The metaphors and the superlatives continue. When the Argentine ant *Linepithema humile* was accidentally introduced from South America into Europe, North America, Australia, Japan and South Africa during the 20th century, lack of genetic diversity in the first colonists meant that what might appear to be neighbouring colonies are so closely related that instead of defensive aggression, the ants simply recognise each other as potential nest-mates, mingling and uniting to form 'supercolonies'. The main supercolony in Europe extends along 6,000 kilometres of coast from Portugal and Spain to France and Italy (Giraud *et al.* 2002), and is thought to contain untold billions of ants. That is an ant nation.

It's easy to slip into nationalist or jingoistic jargon here. Many of the most important human–ant interactions are with invasive alien species that humans have accidentally transported around the world, and which have got out of hand in their new environment. We're relatively lucky here in moist, temperate, northern Europe, because exotic species find the climate challenging, to say the least. Even so, the pharaoh ant *Monomorium pharaonis* regularly grabs the tabloid headlines because of its infestations inside blocks of flats, schools, offices, and especially hospitals. Hidden in the construction fabric of the building, nests of many hundreds of thousands of individuals are tantalisingly out of reach of eradication measures, yet trails of the tiny insects continue to climb down walls, through contaminated rubbish, over food, and over the dressings of vulnerable patients. Towards the end of the 20th century, Edwards & Baker (1981) reported that infestations occurred in about 10% of English hospitals, throughout the country. Two decades into the 21st century and news outlets continue to report 'invasions' of 'indestructible' ants surviving deep inside cavity walls, lift shafts, roof voids and cracked foundations, despite dozens of visits from pest control companies.

Elsewhere in the world, the idea of ants being like invading armies or conquering nations is deep-set. When red fire ants *Solenopsis invicta* were accidentally introduced into the United States from their native South America in the 1930s, they quickly spread in a new landscape which held none of their usual predators, parasites or other natural limiting factors to keep them in check, and they spread like, well, wildfire. Beyond control, these foreign invaders are still berated, cursed and blamed for all manner of economic, domestic, medical and agricultural problems. They violate lawns and patios, interrupting quiet family get-togethers and children's games; they come indoors and infest air-conditioning systems and electrical appliances, as well as foraging in kitchen cabinets and larders. They invade farmland, attacking crop plants and making large mound-nests in cattle fields, then they sting the cows. If they think they are being attacked by animal or human they bite and sting, causing painful welts and red pustules on the skin. What I thought were pins and needles in my legs as I strained to change a flat tyre on the verge of a Florida highway I very soon realised were the burning attacks of these small ants

The pharaoh ant *Monomorium pharaonis*. This species may be tiny, but dense, close-formation trails of this ant can be unnerving and unhygienic.

climbing up my trousers out of the soil where I was standing. It made the jacking and torque-wrench bolting of the wheel quite a palaver, and I probably looked like a demented disco dancer as I hopped and jumped my way up and down the freeway verge to avoid them. By contrast, I've never been stung by any of the closely related *Myrmica* species in the British Isles, even though I've often sat down for my packed lunch nearly right on top of them. It cost me a dollar and a dime for a tube of antihistamine cream to soothe my calves that day, but the national cost of trying to control these alien invader ants in the US is estimated to be $5 billion a year. $5 billion! These are 'just' little ants, right? And it's too late to build a wall to keep them out. They may be small, but ants have big impacts and big implications the world over.

Not all ants are born equal

Another consequence of living in the temperate British Isles is that all of our native ant species have within them a certain uniformity of body size and shape. Turning over a stone or excavating a nest mound produces hordes of workers that are, to all intents and purposes, identical. Each nest will also have a queen (sometimes more than one), and males are produced at certain times of year. Exactly which egg produces a male and which a female (queen or worker) is down to the peculiar genetic system based on chromosome number explored in Chapter 6, but size is intimately tied in to these three forms, sometimes called castes. Queens are usually larger, workers smaller, with males somewhere in between. There may be some insignificant variation within these groups. In the British Isles the largest size discrepancies between individual workers in a nest probably occur in the large wood ants, *Formica* species, with lengths ranging from 6 to 9mm, but even so they are crucially all of the same shape and body proportions. They are, in the myrmecologist's jargon, monomorphic – having just the one body form. This conformity is replaced, though, in some of the more bizarre and highly complex social structures adopted by exotic species, and it's informative to take a look at some of these species to get a flavour of just how diverse and strange the world ant fauna is.

Perhaps the best studied are those New World leafcutters again. Here, instead of all having similar proportions, workers come in various shapes and sizes (dimorphic, two forms, or polymorphic, multiple forms) according to the tasks they perform in the nest, their sizes often reflecting an extreme division of labour. The size classes (also called castes or sub-castes) are variously described as minors (sometimes minims), medias and majors (although up to seven discrete size groupings are known in some species). The workers vary both in overall size and in relative shape; the largest are not simply bigger versions of the smaller workers, as if they had been enlarged on the photocopier, but have certain parts of their bodies overemphasised. One ant, twice the length of another, may have a head four times the width. This disproportionate body distortion is called allometry. Its control is uncertain, but its process can be observed and measured.

Allometry has a nutritional basis during larvahood as some well-fed larvae grow bigger than others, and continues as a genetically guided over-allocation of those nutrients to specific developing parts of the adult exoskeleton during the chrysalis phase. As with all insects that go through metamorphosis, the larval feeding and growing stage accumulates body mass, but all the parts of the future adult structure (legs, wings, compound eyes, antennae, body segmentation) are kept hormonally restrained in a non-developing embryonic form as groups of cells in flattened circular clusters – the imaginal discs. These take their name from the Latin *imago*, meaning an image, from which we also get imagine and imaginary, for the images seen in the mind's eye. Imago is a peculiar jargon word still used by entomologists to denote the adult insect (as opposed to the larva) – being originally the perfect image of the creature as imagined by God, perhaps. The adjective 'imaginal' is only ever used in connection with these embryonic cell-cluster discs, whereas 'larval' is in wide use for anything connected with the immature stages. During the final larval instar and the pupal stage these imaginal discs grow into the adult body structures; they invaginate and flatten into wings or telescope out into cylindrical legs and antennae. Through allometry some of the imaginal discs grow at different rates, and this growth varies between individuals to produce different-sized and differently proportioned workers.

Head-capsule width has often been the easiest way to determine the discrete worker size classes within an ant colony. Measuring and counting the workers often gives two or three peaks on the head-size/population chart reflecting two (major/minor) or three (major/media/minor) worker sub-castes, but this can vary from a single peak (most British and European species) to almost continuous variation across the colony.

Where extreme allometry occurs, majors are often the fighting force of the nest, with over-produced swollen head and powerful mandibles. The largest major *Atta* leafcutters patrol the forage highways and defend the nest against attack; it is the next size down which are the ones actually doing the cutting and bringing back to the nest. Smaller-headed *Atta* workers snip the large harvested pieces into smaller fragments back in the nest, and the smallest minors manipulate the harvest, tend the queen, or nurse the grubs in the brood chambers, while others remove debris. Twenty-two different social functions have been identified in these leafcutters – tasks at least partly defined by body size and body form.

The greatest differential between workers in the same colony is probably in the south-east Asian marauder ant *Pheidologeton diversus*, where the largest majors have a head width 10 times that of their smallest sisters and a body weight more than 500 times as great – that's chihuahua and mastiff from the same mother. Meanwhile, the greatest disparity between a nutritionally surfeit queen and a deprived worker is possibly seen in the Afrotropical *Carebara vidua*, where the huge female can be 4,000 times as massive as her minuscule daughters.

Extreme size can be a limitation on the ant individual. In the seed-collecting *Solenopsis geminata* from Central and South America, the most massive majors have huge furrowed square heads and short

Super major and minors of *Carebara affinis*. This genus shows some of the greatest disparities between worker castes.

Close-up of the head of the tropical fire ant *Solenopsis geminata* from South America, showing the massive head and heavy seed-grinding mandibles.

blunt jaws that are used for milling seeds into edible meal for the colony. Apart from some ineffective self-grooming there is no other task that this monstrosity can perform in the nest; it is the narrowest behaviour in any social insect.

Although polymorphism is well known and well studied, it is still an unusual evolutionary result, and is only exhibited by about 15% of ant species worldwide (only 2% of the 'primitive' subfamily Ponerinae). It is no surprise (though perhaps it is a disappointment) that our impoverished temperate ant fauna does not display this fascinating phenomenon.

How many ants are there anyway?

Ants are individually tiny, but they have an effect far beyond the measure of the individual organisms themselves, and it is measurements of their colonies that should guide how we see and evaluate them. This is not as simple as it sounds, though. It turns out that it is easier to count the grains of sand on the beach than ants in the jungle.

It's one of those tabloid statistics that you can find endlessly iterated through the media – that there is a greater weight of ants on the planet than there is of humans. This suggestion was first mooted by Hölldobler & Wilson (1994), but it was based on calculations done many years earlier on exactly how many ants there might be in the world, and what they might weigh individually. There was quite a lot of guesswork and extrapolation. It starts with the calculation by Williams (1960) that there are an estimated 10^{18} insects living on planet Earth, and a conservative estimation is that 1% of them (10^{16}) are ants. This compares to about 7.8 billion (7.8×10^9) humans. To try and keep the calculations simple, we have to settle on some average weight for both ants and humans, so we really have to leave larvae out of the equation, and also children. Thus, if each adult ant weighs between 1 and 5mg (let's say 4mg), then maybe their combined weight is equal to the combined weight of all adult human beings. But this doesn't quite add up: $10^{16} \times 4\text{mg} = 4 \times 10^{10}\text{kg}$ (40 million tonnes) of ants does not quite equal the weight of all humans over the age of 15 (5.4 billion people with average weight 62kg), 3.35×10^{11}kg (335 million tonnes). Some rejigging of the numbers gets us a bit closer; if we imagine that humans weigh a bit less each, and that a higher percentage of insects are ants, then the two values do start to approach each other. If we take into account when the calculations were done then perhaps ants equalled humans 20 years ago, or maybe 200. Certainly 2,000 years ago ants outweighed us by a considerable margin. In the end, though, these statistics are little more than sound-bites.

What is clear is that ants are, to quote the choice phrase of Hölldobler & Wilson (1990), the 'dominatrices of the insect fauna'. Even if 10^{16} of them sharing the planet with us is still a bit of a guess, other measurements come easier. Mark-and-recapture methods for estimating populations are well known and there are some robust mathematics behind the calculations (see page 310). Since ants make such obvious and discrete nest mounds, it is quite possible to count some of them, sample them, mark them, weigh them, measure their sizes, then examine the colony a short time later to find the necessary data for extrapolation. Though it is quite an arduous task, whole colonies can be excavated and precise numbers can be counted with some certainty. There is a huge range of colony sizes across different species. The North American *Stigmatomma* (formerly *Amblyopone*) *pallipes* has colonies of about 12 (range 1–35) workers, and in Britain the closely related *Ponera coarctata* rarely gets to 60 workers in a colony. The common and widespread red ant *Myrmica ruginodis* usually has between 300 and 3,000 colony members, and the almost ubiquitous black pavement ant *Lasius niger* can get to about 30,000. Wood ants like *Formica rufa* form some of the largest wild colonies in the British Isles, and counts of their mature nests vary from 100,000 to 400,000. Soon, though, it becomes almost physically impossible to collect and count tens or hundreds of thousands of specimens. And justifying the kill of many thousands of organisms, just to count and document them, comes with its own ethical and moral dilemmas.

Towards the upper end of colony size it becomes difficult to get a grip on the sheer numbers. Marauding African army ants *Dorylus wilverthi* are frequently estimated in the region of 10–20 million in a single bivouac and its associated raiding columns through the forest. These estimates are usually calculations based on counting the number of ants in a measured portion of the colony – a bucketful or a measured quadrat of square ground – then multiplying up a reckoned amount. Most of those multimillion estimates for leafcutters are also

Wood ants *Formica rufa* warm themselves in the early March sunshine, but they avoid the cool shadows from the fence cross-struts.

extrapolations, calculated by multiplying up to account for larger colony size from much smaller nests where it was at least possible to find and collect every single insect. What is clear, though, is that there are an awful lot of ants in the world.

Ants live everywhere

To even the most inexpert naturalist, it is clear that ants live everywhere. Going into a field, walking along a road verge or through a woodland ride, the sweep net (my entomological tool of choice) is always full of ants. A walk in the summer woods shows that almost every bush and tree has a line of ants running up and down the trunk. They scuttle out from every rock moved in the garden, every log rolled, and often every piece of turf dug. But ants are actually not quite so ubiquitous.

Like their close relatives, bees and wasps, ants are thermophilic – warmth-loving. They thrive in the tropics and subtropics, and in many apparently barren desert zones, but in temperate latitudes they start to thin out and a clear preference for warm, well-drained sand or limestone soils is soon apparent. They mostly avoid lush riverbanks,

marshes, bogs and swamps, although *Formica picea* survives in the sphagnum bogs of Dorset and Hampshire where it makes small nests of heaped grass fragments on tussocks of Purple Moor-grass *Molinia caerulea* surrounded by quagmire. It seems to be able to cope with inundation and rapidly moves brood if water levels rise. There are a few other wet/cold-tolerant species, but these are often the exceptions, the oddities, rather than the rule. Across northern Eurasia *Formica uralensis* is well known for inhabiting freezing wet Siberian bogs which often flood the nests, and in the Florida Keys the minute African invasive *Cardiocondyla venustula* frequently nests on beaches and can successfully suffer submergence under salt water for many hours at the highest tides. In the Americas, fire ants, *Solenopsis*, are able to survive severe flooding by clasping together in their thousands, along with a cargo of eggs and brood, to create rafts which can last several weeks until the waters subside. On the whole, though, ants shun cool damp regions and inhabit mainly warm and dry areas.

In Britain and Ireland, at least, most ants show a southern, south-eastern or coastal preference – clear indication that higher annual temperature averages and lower rainfall are important for their survival. Some of our rarest ants only occur in small pockets along

If a nest of the fire ant *Solenopsis invicta* gets flooded out, the ants huddle in floating rafts, anchored to whatever herbage they can find, and wait for the waters to recede.

Ants do not like the rain, as shown by this charming vignette from *Episodes of Insect Life* (Budgen 1850) where a bedraggled anthropomorphised four-legged ant probably regrets not taking an umbrella that day.

the south coast of England. *Strongylognathus testaceus*, *Temnothorax interruptus* and *Solenopsis fugax* are good examples; they seem to be clinging onto British residency by the skin of their mandibles. This is a feature echoed by many British bees and wasps and reflects the need for fair-weather foraging to establish nests and stock them with supplies for the young. Cold damp weather is anathema to ants as well, and although they have less danger of falling raindrops clogging their wings and preventing them from getting about (as do bees and wasps), they shelter from the rain and really only get going when the sun comes out. A similar temperature-led gradient occurs up mountains, and although this effect is difficult to demonstrate on Britain and Ireland's rather low-rise hillocks, a recent study in Switzerland showed that fewer and fewer ant species occurred as the altitude increased from 1,800 to 2,550 metres; the species that did survive higher up were also more subterranean in habit (and thus more insulated) rather than mound-making (Reymond *et al.* 2013).

Having said this, there are several British and Irish ants which deviate from this southern geographical bias. The wood ants *Formica aquilonia* and *F. lugubris* are absent from southern Britain, and both have distinctly northern ranges (though neither occurs in Ireland). Elsewhere across Eurasia they are both considered boreoalpine, northern or mountain species; *F. lugubris* in particular is known for being able to forage in damper weather and at lower temperatures than the default southern species of wood ant *F. rufa*, which doesn't quite make it into Scotland. Likewise the common *F. fusca* is gradually replaced by the morphologically similar *F. lehmani* in the north and west, although there is a broad band of overlap where both species occur together. As with many organisms, ants follow a general rule that the further away they occur from some nominal homeland centre of their distribution, the narrower the ecological niche they inhabit, and the more precise their environmental needs. *F. fusca* occurs everywhere in southern England, but further north in Scotland and west in Ireland it starts to become increasingly coastal – a sure link to some critical temperature boundary.

North/south divide

Something odd is going on with two British wood ants though. Both *Formica exsecta* and *F. sanguinea* have peculiar disjunct distributions. Populations occur in the Scottish Highlands, but also in the lowland heaths of southern England – with nothing in between. This can make understanding tenuous and conservation difficult. In particular, although *F. exsecta* is just about holding on in Scotland it seems now greatly reduced in England, and recent work found only a single colony in Devon. Both species are widespread across Scandinavia and central Europe. One possibility is that, for each species, northern and southern populations resulted from two separate colonisation events at different times after the retreat of the last ice sheet some 12,000 years ago; the current north/south divide may reflect subtly different survivability traits within the shallow gene pool of what are still rather outlier populations on the edges of their broader European ranges.

In the British Isles there are barely even 50 ant species that might properly be considered natives, and only 25 of these get to Scotland (Macdonald 2013). This is a seriously depauperate fauna, even counting an additional 10 that have become naturalised after being introduced from elsewhere in the world. By contrast, there are about 100 species in Poland (Czechowski *et al.* 2002, 2012) and 180 in central and northern Europe (Seifert 2007); in France the number jumps to more than 200 (Blatrix *et al.* 2013), and there are more than 270 in the Iberian Peninsula and probably 600 in Europe as a whole. Ants occur from the Arctic Circle to Tasmania, Tierra del Fuego and South Africa. Worldwide there are something in the order of 12,500 described ant species in about 300 genera, with probably at least double this (possibly much more) remaining to be discovered in the deep dark dank tropics.

That ants are thermophilic is clear from the large number of species occurring in the tropics. In the equatorial rainforests, ants are primarily arboreal, living, nesting, foraging in the canopy, in the living or dead branches and trunks of the trees, or associated with the epiphytes growing on the trees. Only a few specialists descend to the forest floor, which although still warm is also wetter, and in many areas is liable to flooding. Meanwhile, in the hottest and driest parts on the globe, ants are also a prominent feature of desert wildlife. If it gets too hot, they can lower their body temperatures by metabolic or behavioural changes. Desert ants in the genus

Cataglyphis scavenge dead insects during the heat of the day, and can sustain temperatures of 50°C, if only for a short time. Long legs keep them a precious few millimetres above the searing sand, in a layer of air a vital few degrees lower. Rushing about at top speed may also offer them a breeze-cooling effect. To seriously cool down they head back to their underground burrows, where a constant mild temperature allows them to recover.

Going underground

Living underground is also one of the reasons for ant success well into temperate regions. Away from the direct burning heat of the sun the subterranean chambers remain cool at noon in summer, but the thermal insulation of soil, wood mulch or leaf litter means that they retain heat even during the bitter night of winter. Ant architecture means there is always some part of the labyrinthine nests where the temperature is just about right. Regular moving of the brood up or down the nest levels keeps the heart of the ant colony at optimum temperature. Movements are timed to seasonal cycles and also to daily temperature changes. In the British Isles, *Formica* and *Lasius* mounds usually have longer, more gently sloping faces to the south to increase sun exposure (this was reputed to be a crude compass for alpine inhabitants in Europe).

Once established, nests offer fine shelter and flexible living conditions for ants. Britain and Ireland's mild, moist oceanic temperate climate offers a wide range of gentle temperature-adjusted niches, so why do we have so few ant species here? There are more than 80 species in Sweden, even though most of that land mass is much further north than Scotland. It is not just climatological factors that have limited the British ant fauna; it has been the impediment of the English Channel and North Sea in preventing colonisation that has kept our fauna impoverished. There was a window, from about 12,000 years ago, as the Weichselian ice sheet started to withdraw at the end of the last ice age, that plants and animals were able to colonise a tundra and birch-scrub Britain still linked to mainland Europe by the land bridge of Doggerland between East Anglia and the Netherlands. Then suddenly, about 6,000 years ago, rising sea-water levels finally flooded the North Sea and English Channel, and the British Isles became islands separated by, to an ant, a daunting expanse of dangerous water. The Irish ant list is even smaller; only about 20 species managed to cross the land bridge before it was engulfed when the Irish Sea flooded 10,000 years ago. This water has kept colonising insects at bay ever since, but the land connectivity of Scandinavia and Siberia and the greater area of stepping-stone islands between Denmark and Sweden have allowed far greater colonisation rates to boost the ant and other faunas of these Norse lands. Elsewhere in the world water has also been the main barrier to ant (and other insect) movements. There are no native ant species in Iceland, Greenland, the Falkland Islands, South Georgia, or the Polynesian islands east of Tonga. We should be pleased with what we've got.

Ants really are very special

Yes, ants are small, and yes, ants are fragile, and yet they are also highly adaptable, highly mobile, highly numerous, and very obvious in their complex community behaviours, even to the non-expert. Everybody knows an ant if they see one – but unfortunately this familiarity has led to some contemptuous and less than helpful stereotypes, and much silly myth and misunderstanding.

What is the point of ants? Do ants bite? Do they sting? Do they farm? Do they harvest? Do they rob? Are they genius architects? Are they nuisance underminers? Do they swarm? Do they go to war? Do they spread disease? Do they destroy crops and gardens? How do they learn? Can they be trained? Is their government democratically run for the colony? Do they live under the shadow of a tyrannical monarch? How long do they live? Will they take over, come the apocalyptic end of the world?

Science is not about knowing all the right answers, it is about asking the right questions. These are some of the right questions to start with. Ants are special because they and their complex colonies are amenable to scientific interrogation beyond that offered by most other insect groups. Though small, ants can give insights into the much bigger picture of all living creatures on this planet – how they interact and intersect with each other and with us humans. The rest of this book details what ants are, how they live, their behaviour and ecology, their physiology and evolution. And although I have tried to report this information in a neutral detached scientific voice, I make no apology for the tone of awe and wonder that frequently overcomes me.

What is an ant? chapter two

The most obvious thing about ants is that they are rather small and ant-shaped. They are, after all, contained within just a single family – the Formicidae – in the otherwise huge and hyperdiverse insect order Hymenoptera, which also includes bees, bumbles, yellow-jackets, hornets, ichneumons, gall-wasps, rubytails, potter and mason wasps, horntails, sawflies and a whole lot more. Ants have a very distinctive look about them – large triangular, square or pentagonal head, slim thorax, narrow waist, bulbous abdomen, long thin legs and elbowed antennae. Sizes and proportions may vary, but this simple formula easily defines ants across the world. Indeed, so familiar is this body form that it is easily stylised in cartoons, logos and graphic motifs. A few simple brush-strokes and the artist has eloquently stated 'ant' for all to see.

In Disney's *A Bug's Life* (1998) the lead characters, though heavily anthropomorphised, are quite obviously ants despite having only four limbs, and coming in various shades of sky-blue. By coincidence *Antz*, the Dreamworks film of the same year, also features the ups and downs of subterranean colony life. The protagonists this time are still simplified, but closer to anatomical and behavioural correctness with six legs (well, four legs and two arms), red body colour, a soldier caste, wars with termites, and jokes about drinking fluid from aphids' bottoms.

On the other hand, with ants looking so similar, identifying them, one species from another, is sometimes a bit tricky, especially from a grainy low-resolution photograph. 'It's an ant' is an easy first-stage identification to make, but going any further is very often just guesswork. As with most insect studies, it is under the microscope, or the hand lens, that the subtle beauty and diversity of ants are revealed. And it is under the microscope that the various anatomical structures of an ant body can be named and recognised.

OPPOSITE PAGE:
Winged or not, there is something easily recognisable about an ant – mostly that broad head and those elbowed antennae. Here a *Myrmica* queen readies for flight.

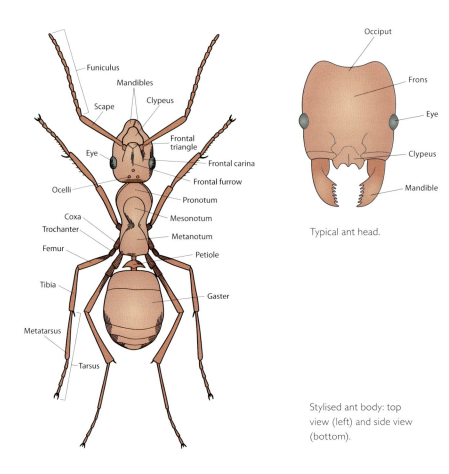

Occiput

Frons

Eye

Clypeus

Mandible

Typical ant head.

Funiculus

Mandibles

Scape

Clypeus

Frontal triangle

Frontal carina

Eye

Frontal furrow

Ocelli

Pronotum

Coxa

Mesonotum

Trochanter

Metanotum

Femur

Petiole

Tibia

Gaster

Metatarsus

Tarsus

Stylised ant body: top view (left) and side view (bottom).

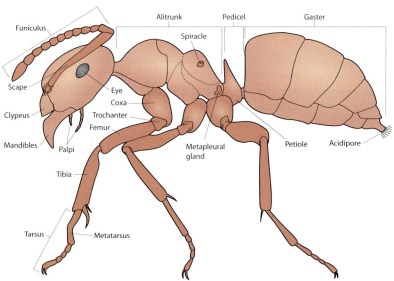

Funiculus

Alitrunk

Pedicel

Gaster

Spiracle

Scape

Eye

Coxa

Clypeus

Trochanter

Femur

Mandibles

Palpi

Metapleural gland

Petiole

Acidipore

Tibia

Tarsus

Metatarsus

The head – brain or brawn?

The head does contain the ant's brain, but at one-tenth of a cubic millimetre this is a tiny organ in a significantly large body part. The small size of the brain is at odds with the seemingly complex behaviour ants exhibit. Charles Darwin (1871) expressed it ably: 'their cerebral ganglia are not so large as a quarter of a small pin's head. Under this point of view, the brain of an ant is one of the most marvellous atoms of matter in the world.' But behaviour, it turns out, is more to do with neural programming than brain size. Rather than storing the intellect, the prime purpose of the broad ant head is to house the musculature which operates the ant's jaws, or mandibles. In most British and Irish species, ant jaws are broadly curved, subtriangular, with a stoutly serrated cutting edge; the jaws meet each other and cross, only just though, and operate more like the jaws of a pair of pliers. Thus it is that ants can carry huge weights in their mandibles, often many times heavier than their own body mass – seeds, stalks, cut leaves, prey, soil particles, their own brood, each other. This precise gripping facility, giving gentle dexterity and also powerful closing action backed up by massive muscles in the head capsule, is often postulated as one of the main reasons that complex

Typical ant form, showing broadly triangular head, narrow thorax, slim waist of just a single segment and bulbous abdomen, together with elbowed antennae and long slim legs. A *Formica rufa* worker exploring.

ant behaviours have evolved – the insect equivalent of the human clenching hand and opposable fingers and thumb.

In the British Isles, only *Strongylognathus testaceus* has long, curved, scimitar-shaped, pranging jaws. This weaponised form is typical of slave-maker ant species (so called because of their raids into the nests of other ants to capture pupae to rear as interned workers; see page 190). It also has slightly enlarged temples – swollen mounds on the hind margin of the head behind the eye to accommodate the extra muscles operating the jaws. This head shape is characteristic of predatory and slave-making ants around the world. Other specialised jaw shapes are possessed by the largest workers of seed-feeding groups like *Acanthomyrmex* and some *Pheidole* and *Solenopsis* species, where the biggest workers are virtually walking heads. The powerful musculature around the broad blunt jaws allows them to efficiently strip away the outer coat of the seeds and grind the endosperm. None of these seed-millers occurs in the British Isles, but any visit to the Mediterranean or Middle East will reveal the famous *Messor barbarus* workers diligently harvesting large seeds and carrying them back to their underground granary silos and pantries.

Very long mandibles, nearly straight with barbed blades or a sudden incurved toothed tip, occur in several exotic genera (e.g. *Anochetus*, *Daceton*, *Myrmoteras*) across most subfamilies, and famously in the pan-tropical genus *Odontomachus*. In this group of so-called trap-door ants the jaws are held open by a locking mechanism at 180 degrees, but are triggered to snap suddenly shut by sensory hairs on the inside of the mandibles, pranging the prey. The Central American *O. bauri* is famously touted as having the fastest strike appendages of any animal, snapping its jaws shut in just 0.13ms ($1/750$ of the blink of an eye), with closing speeds of 35–64m/s recorded and a strike force equivalent to 300 times the insect's body weight (Patek *et al.* 2006). The snap is so sudden that if the ants do not latch onto a substantial anchoring prey victim, the ant is twanged, spinning, 10–15cm into the air by the sudden 10^5g force. Snap-jawing also appears to be deliberately triggered, by striking the leaf or ground surface, to escape enemies and repel intruders. The adventive Australasian species *Strumigenys perplexa* has recently turned up in the Channel Islands. It is tiny, 2mm, and uses its trap-door mandibles to attack springtails.

According to several YouTube channels, large soldier ants of various species around the world are also called surgery ants, because their jaws are powerful enough to close up cuts and wounds in the skin, a

Milliseconds before a mandible strike, a Central American trap-jaw ant, *Odontomachus bauri*, closes in on its cricket nymph prey.

kind of natural medical suturing. Many of the comments below the video clips express doubt and suspicion, no open wounds are shown, and no blood-stemming jaw-clenching skin closures feature; it is often claimed that these are media-stoked urban myths. I'd like to offer some light in the darkness, but unfortunately my anecdote is clouded by the fact that I cannot remember for the life of me who told me their first-hand account many years ago. It runs, in slightly paraphrased form, as follows:

> *I was watching a trail of large ants running across the ground in _____ [some Middle Eastern or North African country] when a local came over to see what I was doing. On seeing my interest in the insects he drew out a large razor-sharp dagger. He then drew the blade across the back of his arm, where it cut a bleeding gash into his flesh. Picking up one of the large ants, he pressed its head against the wound, where it clamped its jaws shut, pulling the edges of the cut closed. Nipping and twisting the ant, he broke off its body, leaving the head, with its deeply embedded jaws, attached to his arm. He repeated this a couple more times until the dagger cut was shut and the blood flow staunched. He then gave a great grin and sauntered away.*

I'd like to apologise to whoever it was who told me this, because I have forgotten who you are. If you're reading this, and feel affronted at my ignorance, please let me know and I'll certainly make amendments

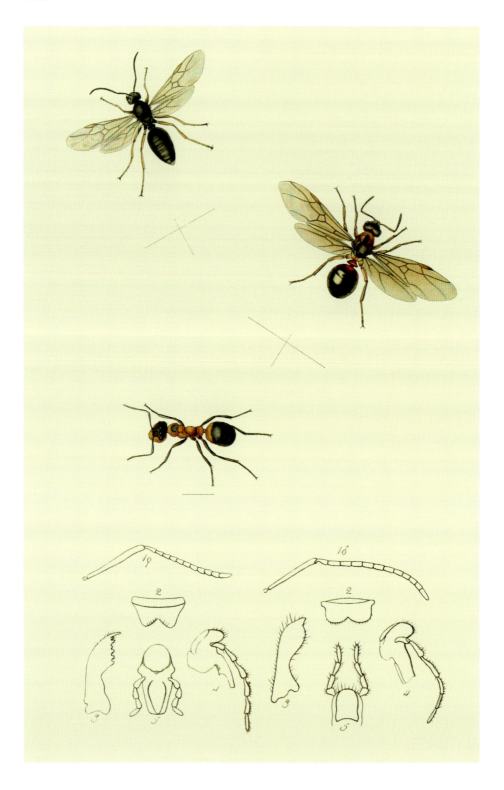

and give full credit and acknowledgement if any second edition of the book ever appears. Meanwhile, a search of the literature does show that this notion of ant-jaw wound stitching is at least well established if not quite conventional. In their commentaries on the history of surgery, Goldenberg (1959) and Schiappa & Van Hee (2012) make at least passing reference to ant-jaw suturing of minor wounds in modern and historical times.

A protective plate over the jaws, the clypeus, covers the mouth and mandibular articulations. Roughly triangular or trapezoidal, its shape and sculpture can offer good identification characters. It forms the leading edge of the typical ant face, and on cartoon ants ought to be where the moustache appears.

Most ant eyes are relatively small. They work in the same way as other insect eyes. There is a variable number of individual lenses, the tiny roughly hexagonal facets visible under the microscope, each of which focuses light down onto a narrow light-sensitive column beneath, stimulation of which triggers nerve impulses for the brain to analyse. These columns (ommatidia, singular ommatidium) combine to form a compound structure – the domed compound eye – but even in dragonflies, with the largest eyes in the insect world, any world-view they produce must be coarse and highly pixelated. In British ants, the eyes vary from a mere handful of ommatidia in most small species (only one in *Hypoponera punctatissima* workers), to compact oval or kidney-shaped organs in wood ants (600 ommatidia in *Formica pratensis* workers), but rarely approach the face-enveloping orbs that can be found in their closest relatives – the bees and wasps. However, like male bees and wasps, male ants generally have larger and more convex eyes than females of the same species (*F. pratensis* again, 830 facets in queens, 1,200 in males). The optic nerve and associated visual centre of the brain are also much larger in males – 70 times the mass of that found in workers of the same species. The traditional explanation for this disparity is that males need to see better to find airborne females at mating time during the well-known flying ant days. There is also some evidence that day-active above-ground hunting and foraging ants have large eyes to visually analyse their surroundings, while fugitive subterranean species have reduced eyes, or none.

Worldwide, the largest ant eyes are found in the bizarre, rare and poorly understood African *Santschiella kohli*, which has massive broad eyes that cover most of the front and sides of the small square

OPPOSITE PAGE:
Exquisite hand-coloured plate from Curtis's *British Entomology* (1823–1840), showing the three castes of the red wood ant *Formica rufa*. From top to bottom: male, queen, worker. The line diagrams below show worker (left) and male (right) antennae and mouthpart dissections.

Worker of the Amazonian *Gigantiops destructor*, showing the huge eyes; those of the male are even larger.

head, although the South American *Gigantiops destructor* comes a close second. Apparently *Gigantiops* is notoriously difficult to study because it can see a myrmecologist coming a mile off and quickly runs away. Workers of many fully subterranean ants (for example African *Apomyrma*, Old World *Leptanilla* and South American *Leptanilloides*) completely lack eyes; so too do the hordes of polymorphic workers making up the raiding columns of *Dorylus* army ants – it seems their concerted military actions rely on their blind faith in tactile and chemical signals, rather than seeing where they are going.

The most characteristic feature of an ant's head is the pair of elbowed antennae – termed geniculate in entomological jargon, from Latin *genu* meaning knee (see fact box – *Geniculation*). Articulated from ball-and-socket joints near the centre of the face, each antenna comprises a long slender basal segment (the scape) and a series of small cylinder- or bead-like segments which together form the flagellum or funiculus (from diminutive forms of *flagrum* and *funis*, Latin for whip and rope respectively). This combination of long solid scape and equally long but flexible flagellum allows ants the joint benefits of broad sweeping sensory arcs as they explore their environment and also close tactile and olfactory sensitivity when examining nest-mates and exchanging liquid food mouth-to-mouth. Male ants usually have shorter scapes than females of the same species, although the flagellum may be longer and thinner.

Typical geniculate ant antenna.

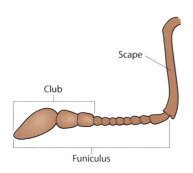

Scape

Club

Funiculus

Geniculation

Geniculate antennae, with a longer first segment (scape), have evolved separately in a few other groups of insects, notably beetles such as stag beetles (family Lucanidae), clown beetles (family Histeridae), burying beetles (subfamily Nicrophorinae), some rove beetles (subfamily Xantholininae) and some chafers (subfamily Cetoniinae). The most geniculate of beetle antennae, though, occur in the weevils (superfamily Curculionoidea).

The weevils are a large hyperdiverse group of mostly plant-feeding species with small short jaws attached to the very front of a distinctive long snout jutting out at the front of the head – the rostrum. This allows the female weevil to drill a long narrow bore-hole into a plant stem, fruit, bud or seed, then turn round and lay an egg deep inside the plant tissue using a telescopic egg-laying tube.

The longest snouts are accompanied by the longest scapes and are well exemplified in the acorn and hazelnut weevils of the genus *Curculio*. As the beetle chews deep into the developing nut the long scape allows each antenna to be flexed back alongside the rostrum

Curculio weevil, showing geniculate antennae.

in special grooves called scrobes, but the apical segments of the flexible hinged funiculus can continue to test and monitor the food substrate all the while. Just as in ants, the scape/funiculus arrangement allows complete manipulative flexibility in the antenna, during both long-distance sensing and close-contact tasting, even when the beetle is face-deep in its food.

In effect, this is the same advantage of flexible manoeuvrability conferred on humans with their elbowed arms – how else would you crash your way through the undergrowth fending off briars and feeling your way through low-slung branches, then later pick up a cup of celebratory tea and deliver it to your lips?

The default ant antenna has 12 segments (antennomeres) in the female (the scape, plus 11 in the funiculus), but 13 (an extra funicular antennomere) in the male. This is usually the easiest way to distinguish winged males and queens in the majority of species. This number has become reduced in a few species, notably to 11 and 12 segments in females and males of *Temnothorax nylanderi* and 10 and 12 in the rare *Solenopsis fugax* and the nearly-British (Channel Islands only) *Plagiolepis taurica*. A few exotic oddities have fewer – South American *Allomerus septemarticulatus* down to seven or occasionally six segments, and the Australian *Colobostruma cerornata* with only four. A general rule seems to be that ants living actively above ground have longer antennae than those living completely subterranean lives, but though there are plenty of ants across the world which completely lack eyes, none ever lacks antennae.

The top of the head is marked with three ocelli (singular ocellus), sometimes called simple eyes to distinguish them from the two large multifaceted compound eyes. These small simple lens structures also occur prominently in bees and wasps, and other insect groups like grasshoppers and crickets (Orthoptera), dragonflies and damselflies (Odonata) and flies (Diptera). Their function is still poorly understood. If compound eyes give a grainy highly pixelated view of the world, ocelli simply register light levels. However, the combination of a relatively large lens and a small aperture beneath it gives a light sensitivity beyond that of the individual or combined ommatidia of the compound eyes. The equilateral or slightly isosceles arrangement of the ocelli occurs across almost all insect orders where they appear, and consensus has it that they aid horizontal level flight by measuring subtle light changes if the insect pitches, lists, skews or yaws during aerial movement – a bit like the artificial horizon monitor on the flight controls of an aircraft. This is supported by the fact that in many ant species the non-flying workers usually lack ocelli, or they are greatly reduced in size. Males of even very small species like *Myrmecina graminicola* also have prominent ocelli, reaffirming their importance during aerial manoeuvres.

In British species the head is always very conventional in form, usually triangular, square or rectangular. Across the globe, however, some very strange head shapes, often in conjunction with jaw shape, have evolved to fit specialist functions and behaviours. As noted above, massive heads, often with swollen occipital margins, contain musculature to power crushing and biting jaws. In the genera *Colobopsis*, *Zacryptocerus* and the aptly named *Cephalotes*, the largest soldiers have the front or top of the head developed into a broad flat rounded or squarish horny plate expanded over the antennae. This exactly fits the size of the small entrance holes into the log-bound colony, and the soldier ants act as living gatekeepers, blocking the tiny nest entrances and only admitting their sisters past their door-shaped heads. In the South American *Cephalotes clypeatus*, sometimes called the turtle ant, the head is expanded at the sides into broad, flat, rounded, nearly transparent plates; these are echoed by similar

OPPOSITE PAGE:
Selection of typical British ants. **I** *Myrmica rubra*, showing the double-segmented pedicel and sharp propodeal spines. **2** *Lasius flavus*, the single petiole plate almost invisible. **3** *L. fuliginosus*, showing the large heart-shaped head and glossy body. **4** *Formica cunicularia*, showing the single-segment petiole plate and two-coloured body. **5** *F. sanguinea*, showing the distinctive notch in the clypeus. **6** *L. niger*, the bog-standard black British ant.

transparent flanges along the sides of the other body segments. This ant is completely arboreal, and the flattened edges to the head and body are probably an adaptation to accidentally falling out of trees – the flanges are aerodynamic, and by twisting its head and body and using its legs like a sky-diver, a falling ant is able to glide, and steer that glide, back towards the trunk of the tree, landing safely rather than tumbling right down to the forest floor (check out YouTube videos of this beast – amazing). The blind *Apomyrma*, *Leptanilla* and *Leptanilloides* mentioned above are subterranean predators of other ants; they are long and narrowly cylindrical, and have long narrow cylindrical heads quite out of proportion to their bodies, tipped by short stout antennae and small jaws. The head of many of the trap-door ants is produced into a short snout, allowing the menacingly toothed jaws, whether curved or straight, short or long, to hinge widely out at the sides.

Thorax, mesosoma or alitrunk?

The conventional division of insects into head, thorax and abdomen tends to fall down when considering ants. These subdivisions, it turns out, are rather arbitrary human constructs and not always part of the natural anatomy of these insects. The middle part of an ant body, what might in everyday conversation be called the 'thorax' by non-entomologists, actually comprises the thorax and first segment of the abdomen fused together. This distinctive feature, where the first abdominal segment, the propodeum, is united with the true thorax, also occurs in all bees and wasps, and in these insects the correct name for this body portion is the mesosoma (Greek μεσος *mesos*, middle, and σωμα *soma*, body, for the middle body section) or alitrunk (Latin *ala*, wing, and *truncus*, trunk, for the wing-bearing body section). Technically this is contrasted with the prosoma (the head) and metasoma (abdomen), but these terms are now rarely used outside technical insect monographs. Many zoologists regularly use the term mesosoma, but myrmecologists have traditionally favoured alitrunk.

Side view of typical ant mesosoma/alitrunk.

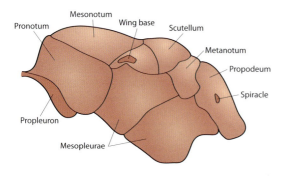

Pronotum, Mesonotum, Wing base, Scutellum, Metanotum, Propodeum, Spiracle, Propleuron, Mesopleurae

The misplacing of an abdominal segment started way back in evolutionary time, so it's important to look back at how invertebrate segmentation arose in the first place. There is precious little in the fossil record, but embryological and genetic evidence indicates that near the dawn of life some primordial, probably soft-bodied creature in the antediluvian murk somehow duplicated or replicated its simple body form into multiple, perhaps self-budding, repeated segments – a process called metamerism. Quite what this progenitor arthropod-type animal might have been is open to far too much conjecture to go into here, but in their magnum opus on the *Evolution of the Insects*, Grimaldi & Engel (2005) offer a masterful synopsis of possible routes and mechanisms. Various suggestions that insect ancestors might have looked like velvet worms (phylum Onychophora), water bears (phylum Tardigrada) centipedes (class Chilopoda) or segmented worms (phylum Annelida) may be flawed, but at least offer a picture for the mind's eye. Segmentation appears to have conferred many advantages and huge bodily complexity to metameric organisms, which vary from obviously segmented millipedes and insects to less obviously segmented humans (think vertebrae and six-packs). Indeed they have, quite literally, inherited the Earth.

Once multiple body segmentation arrived on the evolutionary scene, another process of consolidating these repeated simple segments into specialist groups has given rise to the conventional insect body three-unit quota of head, thorax and abdomen. This is called tagmosis, and a tagma (plural tagmata) is the name of such an integrated body part. Six segments combined to form the head tagma of some long distant insect ancestor, and the relic vestigial limbs are recognisable in the antennae, palps and mouthparts that still occur on the head capsule of modern insects. Three segments combined to form the thorax – technically prothoracic, mesothoracic and metathoracic segments, from front to back. These are readily identifiable today for having a pair of legs attached underneath each, and usually a pair of wings on the second and third segments. Eleven segments are generally reckoned to have been combined to give the abdomen of modern insects. But in the bees, wasps and ants something different has happened, and during the evolution of these groups the anterior segment of the abdomen became closely combined with the three thoracic segments to form the alitrunk.

This fusion is likely to have taken place later in the insects' evolutionary history, rather than earlier. Sawflies (suborder Symphyta) are often regarded as one of the most ancestral groups of the Hymenoptera; they have free-living plant-feeding caterpillar-like larvae and the adults are readily distinguished by the lack of any constricted 'wasp waist' between thorax and abdomen. Although the first abdominal segment broadly joins the third thoracic segment, there is clear division and no amalgamation has taken place. The conjoining of the propodeum onto the metathoracic segment to form the alitrunk seems to be a much later development, and more related to the appearance of the flexible hinge waist so characteristic of ants, bees and wasps.

This ancestral structure of the four mesosoma segments is evident in modern ants, and individual body armour plates can still be identified – and are sometimes important in identification keys. At its simplest, each segment can be envisaged as evolving from a simple four-sided box-kite construction with upper surface (notum or scutum), sides (pleura, singular pleuron) and under surface (sternum). These body plates are most evident in winged queens and males, but are often unclear, merged or poorly developed in the smaller workers.

In ants, the propodeum is a highly distinct section of the alitrunk. On each side is the orifice of the breathing pore, the spiracle; its shape (round or slit) and position (high or low on the propodeal profile) is an important character which distinguishes between the large genera *Lasius* and *Formica*. The two upper hind corners of the propodeum are armed with shorter teeth or longer spines in the vast majority of species in the large subfamily Myrmicinae. Near the lower corner of the propodeum, where it meets the petiole and first rear leg segment, a small orifice shows the position of the metapleural gland. This is unique to ants, and although it has been lost several times through evolutionary history it is regarded as an ancestral trait to identify putative ant fossils. One of its key products is phenylacetic acid (1.4µg, stored in an internal reservoir, about 0.05% of the body weight of a leafcutter worker), which has antibacterial and fungicide properties. Its chemical secretions are thought to perform a variety of tasks from antisepsis to colony recognition and way-marking.

Beneath the mesosoma, ant legs, six of them in the usual full insect complement, are very straightforward. They retain the standard structure typical across other insect groups in having the conventional segmentation, consisting (from body outwards) of coxa (plural coxae), trochanter, femur (femora), tibia (tibiae), tarsus (tarsi; five segments or tarsomeres) and claws. Most ant legs are slim – there are none which have evolved to become fossorial (broadened and flattened for digging, like a dung beetle) or raptorial (spiny for grasping prey, like a praying mantis), as there have been in some wasps. About the only decoration on an ant leg is the spine at the end of the tibia. Those on the front legs are used for grooming the antennae, which are drawn through the fork between the first tarsal segment and the tibial spine. These areas are covered with microscopic combs and brushes of bristles and hairs to keep the antennomeres clean of dust and dirt. In bees and wasps, the middle and hind legs are often used to groom the wings, but if this behaviour occurs in ants it is seldom seen, rarely reported and probably not very important since the flying queens quickly lose their wings anyway and males are extremely short-lived.

Purely decorative, but pleasingly anatomically accurate pavement art, Goose Green, East Dulwich.

Leg length, like antennal length, seems to adhere to the general rule that active above-ground species run about fast on their long shanks, but more subterranean species are slow-moving on shorter stouter limbs. In Britain and Ireland, wood ants (*Formica* species) are extremely rapid in their gait, and a frequent hazard of idly watching their nests is that they are suddenly crawling up your trouser legs, even up onto your head. This can be quite disconcerting for the novice ant-watcher. The weaver ants (*Oecophylla* species) of Africa, Asia and Australia are beautifully and elegantly long-legged, a feature they exploit when making their bodies into living scaffolds and chains to pull leaves together ready for weaving when nest-building. *Cataglyphis bombycina*, the Saharan silver ant, has recently been mooted as being the fastest ant on Earth, clocking up speeds of 0.855m/s. These ants are travelling at more than 100 body lengths per second, and with leg frequencies reaching 40 strides per second (Pfeffer *et al.* 2019). Short-legged ants, on the other hand, rely on stealth and careful deliberate movement through the leaf-litter layer.

The extremely heat-tolerant desert ant *Cataglyphis bicolor*, this one from the United Arab Emirates, holds its body as far away from the searing sand as it can on its long legs.

The appearance of flight in insects is still a hotly debated topic, with several theories as to how and why broad flat aerofoil blades could have evolved. To some extent we can skip past this knotty problem here, since all ants' wasp-like ancestors already had wings. Modern ant wings are remarkably similar to modern wasp wings in their shape, size, wing venation and structure. Lacking the multicoloured scales of butterflies and moths, most ant wings are membranous and clear; they may have some vague clouding in patches, but they lack any strong pigmented patterns. The ancestral insect wing appears to have had eight more or less parallel rib-like veins, analogous to umbrella struts, running from base to apex – precosta, costa, subcosta, radius, media, cubitus, anal and jugal. The first two are fused to create a stiff heavy strut along the leading edge of the wing in almost every extant insect group, and the jugal vein is gone from modern ants. And although these have become fused or partially conjoined over evolutionary time to produce the typical open cell formation and cross-veins visible in wing maps, the other six are still identifiable. Nomenclature of veins and cells has varied from order to order and textbook to textbook over time, but a modern interpretation, based on the cross-order work by Comstock & Needham (1898, 1899) and later revised by Kukalová-Peck (1978, 1983), is shown here.

The stigma (sometimes pterostigma) is the most prominent feature on any ant wing. This thickened, heavily pigmented tear-shaped blotch is thought to have evolved as a counterbalancing weight in the flimsy and highly flexible wing blade of some ancient precursor insect, and it is still clearly visible in the wings of most other Hymenoptera, as well as in dragonflies, lacewings, many flies and some beetles. Ant hind wings are (like those of wasps and bees) small and have greatly reduced wing veins – perhaps just one enclosed cell. They are latched to the front wings by a row of microscopic hooks (the hamulus) along the leading edge, so that fore and hind wings function as a single unified, more efficient aerofoil. The wings are attached either side of the upper quadrant of the second (meso) and third (meta) thoracic segments, but this mechanism, with associated thoracic plates, is only visible in winged queens (even after they have shed or removed their wings) and males.

Despite the fact that ant wings are often used only once, during the 'nuptial' mating flights on flying ant days, ant flight is remarkably efficient. Large females can carry body-weight loads of 7mg per mg of flight muscle, much higher than the insect average of about 5.5mg/mg, or the non-ant hymenopteran 4.6mg/mg (Helms 2018). Flight is not something done with carefree abandon either; it is not simply a means to go off somewhere to colonise a new site. The first (and most important) part of aerial dispersal is meeting a mate, and this requires airborne coordination. Ants do not fly in strong winds,

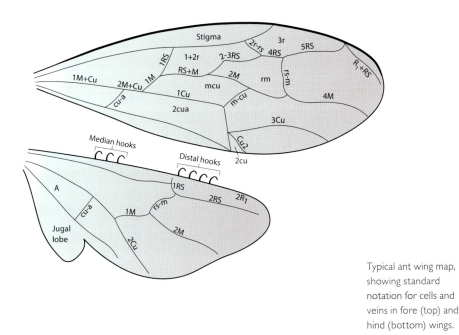

Typical ant wing map, showing standard notation for cells and veins in fore (top) and hind (bottom) wings.

since this would probably just blow them out of the mating arena, and they are loath to take off at air speeds of more than about 1m/s, which is only force 1, 'light air', on the Beaufort scale. This is one of the main reasons that flying ant days only occur on still sultry summer afternoons.

Vein reduction from those original eight struts has been used to confirm the evolutionary classification of modern ants (Brown & Nutting 1949, Perfilieva 2010, 2015) – and ant wings are relatively well preserved in the fossil record. Ants of the subfamily Ponerinae (often described as 'primitive' or 'archaic' for their supposed early evolutionary appearance) show the most complete venation, and have the largest number of enclosed cells, compared to the more 'modern' Myrmicinae, which show several variants by which one or more veins and cells have fused to create a sparser, leaner network (Perfilieva 2010). Care has to be taken, though, because wing size is just as important as phylogeny when it comes to cell number reduction. There appears to be a wing threshold length of 3–5mm under which there are fewer cells. The wings of smaller ants are also (for some as yet undetermined reason) more rounded at the ends.

Wing venation can offer some useful identification pointers for males and queens, but is generally ignored by many field myrmecologists – indeed, the identification handbook by Bolton & Collingwood (1975) contains only two tiny unlabelled thumbnail sketches of wings. This is strikingly at odds with modern identification keys to bees and wasps, which inevitably begin with questions about the wings – the number, shape and size of submarginal cells, curvature of re-entrant veins, size and coloration of the stigma.

Coming to ants through a much wider interest in flies, bees, wasps, indeed all insects, I find wing venation patterns fascinating. Ant wings have a distinct 'look' to them – though Perfilieva (2010) offers 16 different types of ant fore-wing venation. In British and Irish species, at least, the first discoidal cell, slap bang in the centre of the wing, is often square or rhomboidal (but is lacking in *Myrmecina*) and the radio-medial cross-vein drops south from the centre of the stigma, often meeting the median vein to give a strong X (or Greek χ) shape. The common red ant *Myrmica rubra* is peculiar in having the basal portion of the radial sector vein missing, leaving a hanging appendix (abscissa) in the reshuffled cubital cell. Some ant species are particularly prone to venation anomalies, with odd branches and forks spontaneously appearing in random individuals, sometimes asymmetrically in

A queen *Myrmica rubra* suffers the fate of so many winged insects – demise in a spider's web.

only one wing. One suggestion is that more aeronautic species, like *Lasius niger*, which form large aerial mating clouds, are more constrained and precise in the developmental control of their wing venation (any abnormalities are likely to seriously impinge on their ability to fly properly), but those like *Formica rufa*, which have a firmer ground-based meet-and-mate strategy, can afford to have lax genetic control over vein layout, and can therefore get away with more serious vein faults in the wing (Perfilieva 2000).

That worker female ants do not possess wings seems to be down to the same nutritionally influenced developmental processes that also deprive them of functional ovaries. This is a key factor in the evolution of the complex social and nesting behaviour in ants, discussed in greater detail in Chapter 4.

Pedicel and gaster – the rest of the abdomen

Even after accepting that the alitrunk is actually a composite body structure – part thorax, part abdomen – the remainder of the ant's body still does not fit that traditional notion of tripartite insect head–thorax–abdomen anatomy. A close look at any ant soon shows that before the bulbous abdomen proper – the gaster – there is at least one, sometimes two, other distinct body sections to consider.

This body subregion is the pedicel (petiolus in some books), and comprises (in British groups at least) one segment in the subfamilies Formicinae, Ponerinae and Dolichoderinae, and two segments in the Myrmicinae. In the Formicinae the pedicel is a short, but broad, high, linearly compressed scale-like node, shaped a bit like a tombstone leaning up against the gaster. In the Myrmicinae the cylindrical or vaguely conical petiole is followed by the nearly cuboid postpetiole. An intermediate construction is found in the Ponerinae, which ostensibly has just a single petiole segment, but in which the first segment of the gaster is differentiated from the remainder of the abdomen by a belt-like constriction – particularly apparent in the male.

Whether one, two or one-and-a-half bead-like sub-segments between alitrunk and gaster, the pedicel serves an important behavioural function – it acts like a universal hinge, allowing the ant to manoeuvre its tail-tip in almost any direction, left, right, upwards, or down under its body between its legs. This is supremely important, because this is where the ant keeps its sting, and part of the efficacy of any weapon is how quickly, easily and flexibly it can be used. The pedicel, particularly the doubly flexible myrmicine pivot, is pre-eminent.

Like their relatives the wasps and bees, ants have a sting consisting of two long narrow blade stylets that saw rapidly backwards and forwards a bit like the double blades of an electric carving knife. Venom is injected from a muscular reservoir storage sac in the gaster, down a central canal, and the whole structure is protected by outer sheaths (gonostyli) to house the delicate hypodermic apparatus when it is not in use. Ant venoms are poorly studied, but as in wasps and bees they are a proteinaceous cocktail mix of neurotoxins to interfere with nerve transmission, histolytic enzymes that rupture blood vessels and destroy blood cells, and histamines that cause inflammation and swelling. In the British Isles we tend to be rather dismissive of ant stings, even with a media culture dominated by tabloid rantings about killer bees, vindictive wasps and invading foreign hornets. This is partly because our ants are mostly small and ineffectual, but also down to a certain degree of luck in the biogeography stakes. As I discovered on that Florida highway, fire ants (*Solenopsis* species) can be extremely painful, even though no bigger than our native *Myrmica* red ants. We should be extremely grateful that we do not have anything like the Central and South American bullet ant *Paraponera clavata*

Biologist Erica Parra carefully examines a bullet ant *Paraponera clavata* at La Selva Biological Station in Costa Rica. This species is reputed to have the worst sting of any ant in the world. The ponerine abdominal constriction is clear.

anywhere near our shores. Reputed to have one of the most painful stings of any insect, it gets its common name from the fact that its sting feels like a bullet wound. In his famous comparative insect pain scale, US entomologist Justin Schmidt awarded it top 4+ rating, describing it as 'pure, intense, brilliant pain, like walking over flaming charcoal with a 3-inch nail embedded in your heal'. At roughly 20mm long it is a big ant, but the similarly sized army ant *Eciton burchellii*, also from South America, scores only a paltry 1.5. Despite my need for soothing antihistamine cream on my wounds, *Solenopsis* fire ants are rated a feeble 1 ('sharp, sudden, mildly alarming, like walking across a shag carpet and reaching for the light switch'), so I'm pleased not to have experienced worse. In the North American *Pogonomyrmex californicus* and *P. maricopa* the sting shaft is barbed, similar to that of a Honeybee *Apis melifera*, and likewise becomes detached, embedded in the skin of a would-be vertebrate attacker, further increasing the pain even if the ant is brushed away, since the venom sac remains attached and continues to pump in the painful poison.

Another reason British ants lack the dangerous reputation of, say, yellow-jacket wasps, is that not all ants sting anyway. The Formicinae, making up nearly half of British species, and *all* of the common large-nesting species (genera *Lasius* and *Formica*), no longer possess penetrating hypodermic stingers. Instead, they have a round hair-fringed pore – the acidopore – through which they squirt formic acid. Formic acid is such an important ant compound that I'd like to make a brief detour to consider it further.

Also called methanoic acid, formic is the simplest carboxylic acid, with a chemical formula HCOOH. It dissolves (more technically it is miscible, forming a homogeneous mixture) in water and in various organic solvents, has density 1.22g/ml, melting point 8.4°C, boiling point 100.8°C, acidity (pK_a) 3.77, and is colourless, but has a strong pungent smell – not unlike vinegar. It takes its name from the fact that it was originally distilled from harvested wood ants *Formica rufa*; one of the earliest reports was in 1670, when English naturalist John Ray (or Wray as he was then calling himself) described various experiments using 'spirits of pismires'. If you ever get your nose close to a wood ant nest when grubbing about on the woodland floor there is a distinct urinous smell, and this is supposedly the origin of this archaic word – pismire – from *piss* and *mire* (Middle English for ant). Ray reported that when the ants crawled over the blue flowers of chicory thrown onto a disturbed nest, the flowers turned pink. He noticed that the ants 'let fall a drop of liquor' and it was this that stained the petals. Crushing the ants and 'rubbing the expressed juyce against the flowers, I find they will be equally stained'. A similar effect happened when 'spirits of salt' (hydrochloric acid) and 'oyl of sulphur' (sulphuric acid) were dropped onto the bruised petals of these and other blue flowers. Ray reported the similarities with acetic acid: 'Pismires distilled by themselves, or with water, yield a spirit like spirit of vinegar'; furthermore, 'lead put into this spirit … maketh a good saccharum saturni [lead acetate]'.

Ray contemplates how easy this acid is to obtain from the animals, compared to 'those acid spirits, which are extracted out of some minerals not without great force of fire'. He concludes: 'I doubt not that this liquor may be of singular use in medicine. Mr Fisher [the local physician and naturalist who first reported the petal colour change to Ray] hath assured me, that himself hath made trial thereof in some diseases with very good success.' In fact formic acid has found little pharmacological uptake during the last 350 years, but is used as a food preservative (officially called E236) and antibacterial agent in livestock feed, as a limescale remover for cleaning toilet bowls, and in the coagulation of latex in the manufacture of natural rubber. Unlike vinegar it is not readily edible, nor digestible; indeed it is one of the toxins produced by the liver in cases of methanol poisoning (the other being the even more dangerous formaldehyde) when desperate souls resort to drinking bottles of methylated spirits.

Although formic acid is also made and used by several other invertebrate groups (e.g. Puss Moth *Cerura vinula* caterpillar, whip scorpion 'vinegarones'), among the ants it is only manufactured by those in the subfamily Formicinae. Here it mostly replaces the many, diverse and complex venom proteins normally stored in the venom sac, and large species may contain 2mg at 60% concentration. Formicine ants are often regarded as being an evolutionarily more advanced group, but a move away from hypodermic injection stinging to spraying a formic acid aerosol might imply that these ants have given up the powerful defence that still evokes fear and more than a little respect for wasps. But in fact their weapon has become even quicker, easier, and more flexible in operation. Now, instead of dangerous close-quarters melee fighting, the formicine ants stand back and blast their enemies with a noxious corrosive substance from a safe distance. The jet of formic acid often reaches 20–30cm, and can be aimed with some considerable accuracy – the ant equivalent of replacing the poisoned dagger with the CS or pepper spray, perhaps.

Formicine ants are thought to be responsible for previously unexplained concentrations of formic acid in the atmosphere above the Amazon rainforests (Graedel & Eisner 1988); an estimated 1 million tonnes of formic acid are generated and released by the

Wood ants *Formica rufa* squirting formic acid jets to defend the nest.

ants globally each year. To get a good idea of how potent is the ants' defence, a simple trick is to brush the top of an active wood ant nest with a long thin plant stalk and watch the concerted attack as hundreds of workers tuck gaster underneath alitrunk and squirt acid jets up into the air. Don't get too close though, and make sure to keep your eyes clear. The vinegary or urinary smell of the formic acid is very distinctive.

The acid can sting if it gets in your eyes or a paper cut, but is moderately harmless on human skin. Instead, wood ants crawling over the back of your hand will attempt to attack you by gripping individual hairs on your skin, and pulling hard, or by getting their jaws around a delicate fold of flesh. It's not exactly painful, but fairly irritating and can be unnerving if there are scores of the insects all chewing away.

Within the true stinging lineages of the Myrmicinae, Ponerinae and other groups, still not every ant possesses a sting. It is an indisputable fact of nature that the sting of all ants (also bees and wasps) is part of the egg-laying equipment, hence only females have stings. It is no matter that workers are sterile, non-reproductive or non-fertile, they are females; though they may have only rudimentary ovaries, they all have stings or acidopores which they use in prey subjugation or nest defence. Males on the other hand are unarmed, lacking venom sac, stinger hypodermic or acidopore. The stinglessness of male honeybees has been known for millennia (although there was muddle about exactly what sex they were until only a few hundred years ago), and they had their own name – drone. It is a great shame that male ants have no special term to define them – suggestions on a postcard please.

The ant gaster contains more than just a venom or formic acid store – although this is a significantly large component of the internal organs. A detailed anatomical tour of the abdomen is perhaps beyond the scope of this book, but it is important to know that the gaster is made up of overlapping plates in various numbers, relics of that ancient metamerism – tergites above and sternites on the underside. If the pedicel is single-segmented (Formicinae, Dolichoderinae, Ponerinae), tergites and sternites number five in the female and six in the male; if the pedicel is double-segmented (Myrmicinae), one less abdominal segment means tergites and sternites number four in the female and five in the male. The plate-like segments are hinged together, with narrow pleat-like folds of thin intersegmental membranes, allowing the gaster some flexibility of movement and expansion.

Apart from the venom storage sac, the gaster also contains three other key elements – the digestive tract (including a bag-like organ analogous to the crop or stomach), the reproductive tract (respectively ovaries and testes in females and males), and an abundance of secretory glands producing the signal chemicals by which any ant society regulates itself.

In the British Isles, a well-fed ant looks very much like any other, and distension of the abdomen is usually insignificant. But a few years ago the internet was awash with photos of an unnamed Indian ant species gorging on different coloured syrups. As the normally invisible transparent intersegmental membranes were stretched, the strongly food-coloured food was clearly visible inside the ants' bodies. It was all very pretty. This gut bloating is taken to the extreme in several desert-dwelling species throughout the world. Certain large workers of *Myrmecocystus* in the western United States and Mexico, *Camponotus inflatus* in Australia, and *Plagiolepis trimeni* in South Africa (and a few others in Australia, New Guinea and New Caledonia) keep accepting liquid food from their nest-mates until they become swollen to the point of immobility, the gaster being many hundreds of times larger than normal, to the size of a grape, 10–15mm across. The result is a honeypot ant – so called because the 'honey' in their storage crop is accumulated nectar or aphid honeydew little changed from the sugar-rich plant sap that passes through their bodies. Traditionally

Honeypot ants, repletes of *Myrmecocystus mexicanus*, suspended from the roof of their nest cavern.

harvested by aboriginal peoples for their sweet taste, the honeypots, strictly called repletes or plerergates, do not move and usually hang from the roofs of the underground nest chambers. Replete appearance is linked to the seasonal availability of desert flowers, and their bodily stores are built up during periods of cooler moister weather. They then act as food- and more importantly water-reservoirs for the other ants in the colony during later periods of drought.

A bloating that is sometimes visible in British ants is the physogastry of mature queens, when the gaster is over-full of maturing eggs and associated fat stores. Queens are in any case larger and fatter than males and workers, but they sometimes expand to the point where they are barely able to move. Among the ants of Britain and Ireland, one of the greatest size disparities between queens and workers occurs in the common black pavement ant *Lasius niger*, and mild physogastry is sometimes observed in this species. Extreme physogastry also occurs in some termite queens, where a tiny head and thorax are perched on a soft maggot-like body as large as my thumb, but queen ants never match this.

What is not an ant?

This, then, is an ant. And whether we regard them as all looking the same, or we delight in examining minute details under the microscope to tell them apart, we've circled back to the stereotype ant form. Part of their success is down to that simplified default body form – a small manoeuvrable torso on six standard uniform legs, precise jaw dexterity to carry and move things easily, no permanent wings to get in the way of a life lived at least partly underground or in the leaf litter. Ants are general-purpose insects – the jeeps of the invertebrate world.

Having come to a conclusion on what an ant is, it is useful to have a quick look at what is not an ant. The evolution of ants and how they are related to wasps and other insects is covered in Chapter 4. That discussion focuses more on the winged adult queen and male forms, since they are the reproductive units carrying the genetic material and passing through evolutionary selection and descent. The appearance of a wingless worker caste can be regarded as a trivial modern adaptation, but the ubiquity of typical wingless worker ants, with their highly functional pared-down body form, everywhere in the environment, is reflected by a huge mixture of invertebrates

1

2

3

4

5

6

that accidentally or 'deliberately' look like them. It can all be quite confounding to the naturalist tyro.

One group to get out of the way first are the termites. Termites were once sometimes called 'white ants'. They are not ants, but the name has stuck and it's sometimes difficult to get off your shoe as you try to tread down this confusing myth. There are, though, some similarities. Termites live in colonies of hundreds up to many millions of small ant-sized wingless workers. They build complex galleried long-lived nests using soil or wood particles cemented together with saliva. Sexual males and females are winged and take part in massed mating flights, after which wings are shed and a wholly subterranean life is adopted. There are different castes of workers ranging from small nurses, through foragers and builders to large soldiers. They exchange food mouth-to-mouth. They forage for food in a well-defined territory, which they defend against attackers, and each other. But termites are not ants. For a start, they are hemimetabolous – they hatch from eggs as miniature adult-like nymphs and progress to adulthood through a series of gradually increasing instars. They do not have true larvae and do not go through that all-important chrysalis metamorphosis that marks out more highly evolved insects (including ants), which are described as holometabolous. Termites do not have geniculate antennae, ocelli (many don't even have eyes),

Worker termites in the nest living up to their white 'ant' nickname.

petiole narrow waist, fused alitrunk, bulbous gaster, sting or formic acid pore. Their four wings are all similar-sized; front and back flap independently in flight rather than being combined into one aerofoil blade by hamulus hooks. The wings have poorly developed primitive fan-like venation rather than the 'advanced' reduced-vein structure seen in the Hymenoptera. They do not operate haplodiploidy sex determination, and instead of the colony being regulated solely by queens, each nest normally has a single hugely distended egg-bloated female and a consort male – the king. It turns out that termites are not ants, but social cockroaches (for a review see Nalepa 2015).

There are no termites native to the British Isles anyway, but in 1994 a colony of *Reticulitermes grassei*, originally from the Iberian Peninsula, was found in a house in north Devon – presumed to have been accidentally brought back in a pot plant from the Canary Islands. It was treated and monitored and another eradication programme attempted when a few stragglers were found in 2010.

Ant-like winglessness has also partially evolved in several other close wasp lines. Velvet 'ants' (Mutillidae – *Mutilla*, *Smicromyrme*, *Myrmosa* in the British Isles) are actually wasps, and have winged wasp-like males, but wingless females. They are certainly velvety, although *Myrmosa atra* is the only British species that really looks anything like an ant; it has a narrow red alitrunk, and the first segment of the gaster is also red, contrasting with the remaining black abdomen in a good mockery of a pedicel node. *Mutilla* is rather stout and tubby, and half resembles a wingless bumblebee. None possesses geniculate antennae. Mutillids are parasitoids in soil-nesting bee and wasp nests, the female sneaking into a target burrow to lay her eggs on the host larvae while the owner is off foraging.

The Bethylidae (no common English name for this family of tiny solitary wasps, I'm afraid) are tiny and slim, and although they are mostly winged, they have a tendency to run about rapidly, rather than fly. They have slim dark shining bodies and elegant pentagonal heads. In the sweep net they look like a cross between ants and tiny rove beetles. Little is known of their biology but most are thought to be parasitoids, attacking larvae of moths or beetles in cryptic situations like leaf rolls or burrows. Their close relatives the Dryinidae have wingless very ant-like females, particularly as the thorax is often constricted in the middle to give the appearance of alitrunk and pedicel node. They are parasitoids of leafhoppers, and are immediately distinguishable under the microscope for having chelate (pincered) front legs for grappling their victims. Again, bethylids and dryinids possess neither true pedicel nor geniculate antennae.

The highly speciose and very diverse ichneumon wasps have many small wingless forms (often the females) scattered through various families including Ichneumonidae (e.g. many common *Gelis* species, parasitoids of various invertebrates including moth caterpillars, spider egg sacs, and other parasitoid wasps), Braconidae (several very obscure genera), Diapriidae (e.g. *Platymischus dilatatus*, a widespread but under-recorded parasitoid of *Orygma luctuosum* flies breeding in strand-line seaweed), Scelionidae (e.g. the rare *Baeus seminulum*, a rare squat rather beetle-like parasitoid of spider eggs), Platygastridae (*Platygaster*, rare parasitoids of leaf-rolling Cecidomyiidae flies), Chalcididae (many obscure parasitoid genera) and Cynipidae (e.g. the common oak-apple gall wasp *Biorhiza pallida*). This last species looks very antish in the size and proportions of head, alitrunk and bulbous shining abdomen; it is exactly the same colour as the widespread tree ant *Lasius brunneus*, and can often be found climbing up the same oak trunks.

Along with ant-like hymenopterans, there is a vast swathe of patently non-ant-related ant mimics among many other groups of invertebrates. A question often asked is whether this mimicry achieves some protection, giving the impression to potential predators that this is a stinging or acid-spraying formicid, rather than a tasty morsel. There are few answers, but the deceptions are often so remarkable in shape, coloration and movement, that we should keep asking, studying and observing.

At the prosaic end of the 'myrmicry' spectrum, ant beetles, family Anthicidae, are small, dark, mostly sombrely patterned, but with broad heads, spindle-shaped pronotum and relatively broad shoulders to the wing-cases. They do, at first sight, look a bit like ants. They are rather secretive, running about on bare ground or through light herbage, but their resemblance could easily be convergent evolution towards the general-purpose body form of the ground beetles, family Carabidae, with which they are also easily confused. *Thanasimus formicarius* (family Cleridae) is sufficiently ant-like to have acquired an appropriately formic scientific name. Its coloration does slightly resemble that of an ant – one with black head, red alitrunk and black gaster. This is a common colour scheme in this relatively scarce beetle family, and although *Thanasimus* is usually found under dead bark, hunting out its wood-boring beetle prey, exotic species occur on flowers

The red base to the wing-cases, edged with chevron bands of pale scales, gives a gaster-like appearance to the longhorn beetle *Anaglyptus mysticus*.

The contrasting red at the centre of its body gives *Thanasimus formicarius* a good ant alitrunk/gaster appearance.

and herbage, and in truth clerids more closely resemble *Mutilla* velvet ants (Mawdsley 1994). Several other unrelated beetles also adopt this black/red/white colour pattern, including in Britain the longhorns *Anaglyptus mysticus* and *Poecilium alni* (family Cerambycidae).

Ant-flies (family Sepsidae) are small, shining black, with large round heads and slightly waisted abdomen before the more bulbous tip. Any resemblance to ants is rather superficial; they breed in animal dung and other decaying organic matter and are frequent in grazing meadows, compost heaps and rotting seaweed.

A remarkable number of the true bugs, Hemiptera, closely resemble ants. Adults of our three local but widespread arboreal *Pilophorus*

A young Lobster Moth *Stauropus fagi* caterpillar showing highly ant-like form with reared 'head' at the tail end.

species (family Miridae) again echo that black/red/black-with-white-bars mutillid colour scheme earlier modelled by *Thanasimus*, as does the rare *Systellonotus triguttatus*. In this last species, the nearly wingless females are narrow-waisted and bulbous-abdomened and look extremely ant-like. Other remarkably anty body forms can be found in the wingless forms (often female) of *Macrodema micropterum* (Lygaeidae), *Loricula elegantula* (Microphysidae), *Mecomma dispar*, *Pithanus maerkelii* and *Myrmecoris gracilis* (Miridae), and the wingless nymphs of *Coreus marginatus* (Coreidae), *Himacerus mirmicoides* (Nabidae), *Alydus calcaratus* (Alydidae) and *Miris striatus* (Miridae). In all these hemipterans, an ant impression is emphasised by a rapid jerking gait. Many early-instar larvae across several other insect orders also resemble ants, notably the new hatchlings of the Lobster Moth *Stauropus fagi*, with its upturned bulbous tail-tip closely resembling an aggressively poised ant head, and across the globe this feature appears in the nymphs of many species of mantis, stick insect and bush-cricket. Early-instar nymphs of the Australian stick insect *Extatosoma tiaratum* resemble (in both form and gait) the 'spider' ant *Leptomyrmex ruficeps*, which occurs alongside it in the same forests.

Other confusingly ant-like forms are also common among spiders. In the genera *Phaeocedus*, *Callilepis*, *Phrurolithus* and *Micaria*, strategically placed patches or bars of white scales on the base of the black abdomen can very convincingly emulate a petiole node. The most strikingly ant-like of British spiders are *Myrmarachne formicaria* and *Synageles venator*, which both have slim colour-emphasised bodies concealing the typical two-segmented spider form (cephalothorax and abdomen) and suggesting the four-segmented ant form of broad dark head, narrow alitrunk, petiole node and bulbous gaster. Combined with rapid ant-like movements and often occurring with ants (on which they prey), they are remarkably well disguised.

These are just the British species. When the horizon is shifted to include worldwide ant mimics some utterly bizarre shapes and forms come into view. The Brazilian spider *Myrmecium latreillei* has its cephalothorax contracted into a delicately shaped alitrunk- and petiole-like nodular form, and its palps are long and angled like geniculate antennae; so close is the formicine approximation that the arachnid's extra pair of legs does nothing to expose the fraud. The Ecuadorian spider *Aphantochilus rogersi* has the size, shape, spines, colour and texture of the dull matt-black turtle ants *Cephalotes atratus* on which it preys – almost an exact match.

In the forests of the Philippines, a slim, dark moth, *Xestocasis iostrota*, holds its wings rolled furled and its spiny legs out sideways, and has its wing-tips dilated into a phantom head; as it scuttles across the leaves – backwards – it looks just like a foraging ant.

One of my favourites is the tiny South American leafhopper *Cyphonia clavata*, which has its pronotum sculpted into an elaborate shining black bobbly likeness of an ant. The odd thing is that the 'ant' is facing backwards over the insect's rear end, with terminal bulbs like large round eyes and long stalks resembling antennae. The base is enlarged into a rounded nodule like a gaster, and in the middle is a spined constriction in exactly the right place for a petiole node. The deceit is remarkable, and this insect has become a minor internet sensation.

The ants of Britain and Ireland

There are currently about 50 species of ant that might be regarded as genuinely British or Irish – these vary from the ubiquitous to some which are perhaps of dubious status here and others that are probably extinct (if they were ever truly established in the first place). In addition there are several that have been intercepted at entry ports, some found temporarily resident in heated buildings (especially glasshouses), and two that were brought into the country on huge palm trees forming part of grand temporary show-garden schemes. A further three species are known from the Channel Islands, and there are several mainland European species that appear to be on the move, and which might conceivably colonise the British Isles some time in the near future.

What follows is not a full formal descriptive catalogue of British and Irish ant species. Such an undertaking goes slightly beyond the remit of this book. Rather, it is an introductory guide to British and other nearly-British species. The identifying distinctions between ants can be rather technical, and can sometimes only be determined by examining set museum specimens under a stereomicroscope. For anyone wishing to go further with their ant-naming skills, there is an illustrated identification key in the Appendix at the end of this book.

Note on scientific names

Following long-held biological convention, the binomial scientific names of insects are based on a genus name, which is capitalised, and a specific name, which is not, and these are usually italicised thus:

Ponera coarctata

OPPOSITE PAGE:
Tetramorium caespitum,
one of the less common
British species.

Also by convention, the name of the first person to formally describe the species in print is given, along with the publication date of that description. Rather abstrusely, the describer's name and date are sometimes (but not always) given in parentheses, and this reflects exactly what the species was originally called at the time. When renowned French entomologist Pierre André Latreille (1762–1833) published his monograph on ants in 1802, he described a new ant species *coarctata* for the first time, but placed it in the only ant genus then available, so it was listed as *Formica coarctata*. The novel genus name *Ponera* was later suggested by Latreille in 1804, and his species became *Ponera coarctata*; his authority was now parenthesised when it was moved to another genus, so:

   Ponera coarctata (Latreille 1802)

Conversely, Italian entomologist Carlo Emery (1848–1925) described a pale yellowish colour variant *Ponera coarctata* form *testacea* as a new ant subspecies in 1895, and it was raised to full distinct species status in 2003; it was, and has remained, in the same genus ever since it was first described, so his name and date do *not* require parentheses, and it remains:

   Ponera testacea Emery 1895

For ease of reading, the formal authority and date notation is often dropped in the running script of books, and for the most part I have done this, to keep my text as free of clutter as I can. However, it is used in checklists, descriptive monographs, official species reports and scientific papers, so in this chapter I have included these details for completeness.

Really, though, this type of information only becomes vital for somebody sorting out misidentifications and misdescriptions during an international review of species. Sometimes it turns out that a single species has been independently described using multiple names by different entomologists, either from different parts of the world or at different times. And conversely, the same name may have been used independently for different species in different parts of the world. Consensus needs to determine how to untangle these muddles for the future, to avoid confusion. In these cases precedence is usually given to the earliest named description, and there are complex rules to interpret these mix-ups, and an international nomenclature commission to adjudicate any conflicts.

Note on common names

Contrary to scientific nomenclature, simple common folk names have been cobbled together down the centuries by simple common folk, who did not have and did not require any detailed understanding of an organism's evolution or taxonomy. This has led to a situation where English names are of dubious and uncertain value when discussing certain groups of organisms. Unlike, say, plants, birds, butterflies and fungi, British ants have never acquired useful common names, because in common conversation even now they rarely need to be identified much beyond 'ant'. Even the frequently used 'black pavement ant' and 'yellow meadow ant' are still vague descriptive forms for the *Lasius niger/L. platythorax* and *L. flavus/L. myops* complexes, rather than fixed common names for particular species. This has put this book at odds with others in the series where English common names are widely accepted and widely used.

A few British ants have recently acquired English names because nature conservation organisations sometimes need to communicate important biodiversity information to a wider public audience who might be put off by unfamiliar scientific terms. When Buglife needed to drum up public sympathy, political awareness and financial support to promote their work on studying and conserving the rare *Formica exsecta*, a new English name needed to be coined, and thus we now have the 'narrow-headed ant'. Such innovations are welcome and worthwhile, but it must be remembered that in the more obscure groups of animals (like ants), scientific and common names do not have parity, they are not equal, and the English names must always take a subsidiary and subordinate role. Consequently, I have resisted capitalising English names of ants, because this would give them a status, which they do not deserve, comparable to their scientific names. I have also resisted creating new English common names for the ant species included in this book, and I use any existing names rather grudgingly.

Note on geographical coverage

According to the dust-jacket notes, the books in the British Wildlife Collection are on all aspects of British natural history. However, wildlife does not respect national or political boundaries. And what does British mean anyway? This is the rationale I have used for this book. Britain, sometimes Great, is the island group that comprises the countries England, Wales and Scotland. Across the Irish Sea, Ireland comprises the Republic of Ireland and Northern Ireland – that part which allies itself with Britain to make up the sovereign country the United Kingdom of Great Britain and Northern Ireland. Together, along with all the outliers, the anomalous Isle of Man in the middle, and sometimes the Channel Islands, these form the British Isles. I acknowledge that this term comes with some historical and political baggage. When referring to the wildlife of these islands I sometimes use 'British and Irish', and at other times just 'British'; the two descriptors are often interchangeable. I apologise that I have sometimes fudged my words a bit in the interests of readability, and for the avoidance of repetition or clumsy constructions.

Note on descriptions

The species descriptions and comments given below are not formally comprehensive, nor are they exhaustive; that would have made them repetitive and unreadable. Instead, I have written more about some important, characteristic, distinctive or charismatic species, and much less about others which are extremely similar. Sizes are head-to-tail body lengths, not including antennae or legs. Dates for the alates (winged queens and males) are broadly the times when they might be expected to be seen on the wing.

Family Formicidae
Subfamily Ponerinae

The Ponerinae are small, secretive, slow-moving cylindrical ants, characterised by having a single-segment pedicel, and an indented constriction between the first and second abdominal segments which makes them look as if they've done their belt up too tight. They are primarily predatory species and make small nests under rocks and logs.

Ponera coarctata (**Latreille 1802**)

Worker 3.0–3.5mm. *Queen* 4.0–4.5mm. *Alates* August and September.

This is a small dark, narrow, secretive, slow-moving ant. It has an elongated abdomen, much less bulbous than many ant gasters, which is very distinctive, and together with the relatively short stout legs and antennae often makes this ant look a bit like a small rove beetle as it scurries off down a crevice in the soil. It usually makes

small colonies of about 40–50 workers with two or three queens, under stones and logs, in moss, or in broken stony ground. This is a predatory ant, feeding on tiny soil invertebrates. It is right on the edge of its European range here, and is more or less confined to warm, sheltered, though not necessarily particularly dry, sites in southern England. For much of its British range it is coastal, with records from Cardiff (the only Welsh record?) to Colchester, but its headquarters seems to be the Surrey heaths, Thames estuary and London basin. It is probably under-recorded and is easily overlooked. When testing a two-stroke garden blower-vac for use as a suction sampler, I tried it out in my back garden and found this species commonly in my disreputably uneven and patchy south London lawn; I've never found it there by fingertip grubbing (gardening) though.

Ponera coarctata, showing its cylindrical form and constriction between first and second abdominal segments.

Ponera testacea Emery 1895

This species is extremely similar to *P. coarctata*, of which it was considered a subspecies until very recently. It is generally pale orange-brown rather than dark brown-black, with subtle differences in the shape and size of the petiole, which is shorter and thicker in *P. testacea*, barely reaching level with the top of the gaster. Its biology is probably similar to *P. coarctata*. It was first noted as British from Dungeness by Attewell *et al.* (2010), and it has since been found in a few coastal sites from Cornwall to Suffolk (Paul 2011). Some of these are modern records, others are redeterminations of specimens previously identified as *P. coarctata*. It may turn out to be more a species of hot dry well-drained soils than its congener. Whether this is a recent colonist to England or has simply been overlooked for decades may have to wait for a retrospective analysis of specimens buried in private and museum collections.

Hypoponera punctatissima (**Roger 1859**)

Worker 2.5–3.2mm. *Queen* 3.5–3.8mm. *Alates* August to September.

In form and behaviour, this ant is very similar to *Ponera coarctata*, and is frequently mistaken for it. Males are wingless and worker-like. It is usually only found inside or associated with heated buildings, particularly hothouses and conservatories, but occasionally turns up well away from human habitation. It is a predatory ant, so its appearances inside buildings are not scavenging forays after human food stores, it is simply that it requires shelter from the extremes of British cold weather to survive. It still needs access to soil-dwelling prey items though. Colonies contain several females and up to about 200 workers. Despite its association with buildings, it has a good claim to being native, and subfossil remains have been found in northern England dating from 1,500 years ago. Although less often found than *Ponera coarctata*, its distribution here is more broadly scattered, with records throughout England, Wales, Scotland and Ireland. I've seen it only once, a lone worker, in Nunhead Cemetery, under the bark of a fallen tree, 23 October 1996 (Jones 1998). Although the cemetery is now a wild and wooded site, it is in the heart of urban south London, and surrounded by houses, schools, shops and offices.

Hypoponera ergatandria (**Forel 1893**) (**formerly** *H. schauinslandi* **Emery 1899**)

This ant is slightly smaller than *H. punctatissima*, but size ranges broadly overlap, and it can only reliably be separated on extremely subtle body measurements of head width and scape length (Seifert 2013). This is a tropical or subtropical species but has spread around the world as a tramp, occurring in heated buildings. It is reputed to turn up in the British Isles very occasionally.

Subfamily Dolichoderinae

The Dolichoderinae have a single-segment pedicel, and rather than a circular acidopore formic acid squirter at the tip of the gaster there is a narrow transverse slit. This is a rather insignificant subfamily in the British Isles, with only two rare native species and a few others that have been introduced from elsewhere in the world that can only survive in heated buildings. It does, however, contain the well-known Argentine ant *Linepithema humile*, which is one of the most important ant species worldwide, and which could potentially become a significant part of the British fauna if climate change continues to facilitate its increase here.

Linepithema humile (Mayr 1868) (formerly sometimes called *Iridomyrmex humilis*), the Argentine ant

Worker 2.2–2.8mm. *Queen* to 4.5mm. *Alates* April to June.

This is a small active pale to dark bronze species. In side view the propodeum is arched, and constricted where it joins the thorax. It is easily mistaken, in size, colour and activity, for the common black pavement ant *Lasius niger*, but it is generally smaller, paler and more slender. A well-known and widely travelled tramp species, it is originally native to South America, and has been accidentally transported around the globe. In Britain and Ireland it sometimes forms colonies in heated buildings or glasshouses. A large outdoor colony was recently found in several streets in Fulham, west London, covering an area of about half a hectare (Fox & Wang 2016). The colony was living in paving and behind brick garden walls, with nest entrances well away from any buildings, suggesting that it can survive out of doors in British cities. It is likely to be discovered more often in the open in future.

The Argentine ant *Linepithema humile*, attending aphids.

Linepithema iniquum (**Mayr 1870**)

Worker 2.0–3.0mm. *Queen* 4.0mm.

A small, dark, active ant, it is distinguished from *L. humile* by the even greater mesonotal impression and step-like bump on the rear of the alitrunk, and by the glossier and less hairy body. In its native South and Central America it is primarily an arboreal species, and many of the intercepts in European ports have been on imported epiphytes. It was reputed to have been recorded in Ireland and has recently been found breeding in the heated tropical houses of the National Botanic Garden of Wales (Hamer & Cocks 2020), though it is thought they had been established there for about 10 years.

Tapinoma erraticum (**Latreille 1798**), **sometimes called erratic ant**

Worker 2.6–4.2mm. *Queen* 4.5–5.5mm. *Alates* June and July.

This dark brown or black ant is highly active and aggressive if disturbed. It slightly raises its gaster threateningly, but does not possess an acidopore. A volatile chemical is sprayed through the slit-like anal vent – described as smelling of rancid butter by Blatrix *et al.* (2013). It is more or less shining, and lacks the body hairs of the common *Lasius niger*, which it superficially resembles. The head is also longer behind the eyes, wedge-shaped, and the frontal triangle is poorly defined. The genus is relatively easily separated by the overhanging first gastral tergite, which projects forwards

Tapinoma erraticum worker with brood.

and obscures the petiole segment. Nests are usually under stones in well-drained rocky or sandy places, and comprise a few hundred workers and numerous queens. This species is mainly carnivorous, but has also been recorded visiting leafhoppers after honeydew. It is a scarce warmth-loving species of dry sandy heathy places, and most British records are from the Dorset, Hampshire and Surrey heaths.

Tapinoma subboreale Seifert 2012 (formerly called *T. ambiguum* Emery 1925, or *T. madeirensis* Forel 1895 by British authors)

Virtually identical to *T. erraticum*, this species has only recently been recognised as British. It is separated by the shape of the clypeus, a character that is difficult to see, very comparative, and rather doubted by Pontin (2005), who suggests that male genitalia should really be examined. Its biology and nesting behaviour are thought to be similar too, but little is yet known of its British distribution or status.

Tapinoma melanocephalum (Fabricius 1793), sometimes called ghost ant
Worker 1.5–2.0mm. *Queen* 2.5mm.

This ant is smaller than *T. erraticum* and has a distinctive bicoloured body: head dark, alitrunk brownish, and gaster pale but sometimes marked with brown blotches. The legs, antennae and mouthparts are very pale. This is a tropical species that has been accidentally transported all around the world for so long that its original native range is not known. It is occasionally introduced into the British Isles with horticultural material, or in food produce, and establishes colonies in hothouses and heated buildings. Perhaps this is one to look out for in the supermarket every so often.

Tapinoma nigerrimum Nylander 1856
Worker 2.5–5.0mm. *Queen* to 5.5mm. *Alates* June, then September and October.

This species is slightly larger and blacker than any other *Tapinoma*. It is a southern European species that has been found introduced into urban areas in Belgium and Germany (though not inside buildings), so might turn up in the British Isles in the near future.

Subfamily Myrmicinae

The Myrmicinae make up the familiar red ants found nesting under rocks and logs everywhere. They are immediately distinctive because the pedicel node consists of two segments – petiole and postpetiole. The rear corners of the alitrunk are usually armed with sharp teeth, sometimes long curved spines. They possess stings, but in most British species these are rather ineffectual as they have neither the length nor the strength to puncture human skin unless they find a soft spot.

Aphaenogaster subterranea (**Latreille 1798**)

Worker 3.0–5.0mm. *Queen* 5.0mm. *Alates* July to September.

This is a bright to dark red-brown ant, with short bulbous gaster and rather hunched thorax. The legs and antennae are yellowish red, and the head, alitrunk and gaster have numerous erect bristles. The short propodeal spines are a key feature. It makes small nests of a few hundred workers and a single queen under logs, mostly in woodlands. Although not recorded from the British Isles, it is widespread on mainland Europe, including France and Belgium, so is a potential coloniser in the near future.

Crematogaster scutellaris (**Olivier 1791**)

Worker 3.5–5.0mm. *Queen* 8.0–9.5mm. *Alates* August to October.

The reddish-orange head contrasts with the brown alitrunk and shining black gaster. The gaster is distinctly heart-shaped – broad and

Workers of *Crematogaster scutellaris*, showing heart-shaped gaster and red head.

rounded at the base, pointed at the tip. The postpetiole articulates with the upper surface of the first gastral tergite and gives the ant the ability to flex its abdomen up over its alitrunk in a menacing threat display if it feels disturbed. Although not a native British ant, this is a common and widespread species in southern Europe where it is mainly tree-dwelling in trunks and logs. Colonies comprise many thousands of workers and a single queen. A non-invasive technique for locating nests is to bang on tree trunks with a large stout stick. Defensive columns of ants pour out of nest holes to investigate the disturbance. This behaviour may account for the ant's occasional introduction into the British Isles in cork shipments from southern Europe or North Africa, as defending hordes become trapped in harvested cork bark to become temporarily established in British warehouses.

Formicoxenus nitidulus (**Nylander 1846**), **often called shining guest ant**

Worker 2.8–3.4mm. *Queen* 3.4–3.6mm. *Alates* July and August.

This is a small shining reddish- or yellowish-brown species, with the gaster often slightly darker. The body has short scattered hairs. The head is rather long and parallel-sided, appearing swollen in front of the small eyes. The top of the alitrunk is rather flat, and the propodeal spines are rather short and blunt. It is an inquiline guest, making small colonies of about 100 workers and several queens (though only one egg-laying) inside mounds of the wood ant *Formica rufa* (also

Beside its *Formica rufa* host, it is obvious how the diminutive *Formicoxenus nitidulus* acquired its some-time English name 'shining guest ant'.

F. aquilonia and *F. lugubris* in Scotland and northern Britain, Macdonald 2013, Robinson 1999, Stockan *et al.* 2017). Several nests, each made inside wood fragments, hollow twigs or stems, may be present in one *F. rufa* mound. The ants are ignored by their hosts, from whom they solicit or steal food. If a large ant approaches, *Formicoxenus* stands still; if it is picked up by a *Formica* worker, it is immediately dropped – chemicals on its cuticle making it distasteful to its hosts. The narrow entrances to the nest-within-a-nest are too small for *Formica* ants to explore. The queens are winged but the males are wingless and closely resemble workers. Although recorded widely in Britain, from Dartmoor to the Trossachs, the only real cluster of records is in the heaths of the Surrey/Hampshire border. It is probably easily overlooked. According to Pontin (2005), one way of finding them is to push old bracken stems into a wood ant nest, and pull them out again several months later to see if *Formecoxenus* has taken up residence inside them. On warm dull days the *Leptothorax*-like workers can sometimes be found walking over the exterior of the host mound, distinctive by virtue of being small and very shiny among their larger duller hosts.

Leptothorax acervorum (Fabricius 1793)

Worker 3.8–4.5mm. *Queen* 3.8–4.8mm. *Alates* July to September.

This species is pale reddish yellow, with the head, antennal club and large first gastral tergite darker. It is generally larger and more

Worker of *Leptothorax acervorum* carrying a piece of dead grass back to the nest.

robust than members of *Temnothorax*, which were once also included in this genus. In wooded areas it normally nests in small logs, hollow tree trunks, stumps and rotten fence posts, but it will also form small colonies of up to about 100 workers, in well-drained soil around plant roots. It is sometimes found nesting next to or on the margins of large *Formica* nests, and although it walks with the larger ants it remains unmolested. This is a relatively common and widely distributed species throughout the British Isles, but is secretive and easily overlooked. This is apparently the only ant species thought to occur naturally throughout the Holarctic – Europe, Asia and North America (Schär *et al.* 2018).

Monomorium pharaonis (Linnaeus 1758), the pharaoh ant

Worker 2.0–2.4mm. *Queen* 4.0–4.8mm. *Alates* all year round.

This tiny ant is pale reddish yellow, though the gaster is obscurely darker in some individuals. It is a slim, elegant species, with rather parallel-sided head and narrow petiole. It is one of the most widespread ant species in the world, having been accidentally transported all around the globe by unsuspecting humans. In the British Isles it is only known inside heated buildings, where the well-hidden nests sometimes contain millions of workers and thousands of queens. It is a particular pest of hospitals, where constant temperatures allow it to keep colonies deep in the structure of the building in hidden corners that are inaccessible to pest controllers.

Monomorium pharaonis workers.

It usually has to be treated with poisoned ant baits. Pharaoh ants scavenge, and send long columns into food areas, and are also blamed for traipsing diseases from contaminated dressings or other clinical waste. As with many domestic insects, this ant is widely reported by pest control companies, but is seldom recorded by field entomologists. However, it is likely to occur in almost any town or city in the British Isles. When I was presented with one from Deptford in 1998 the donor told me that she had found it in one of several local shops where the ants seemed to have free run of the shelves. The similar *M. floricola* (Jerdon 1851) is another tramp species transported widely around the world, but mostly in tropical zones. Its head and gaster are dark, contrasting with the pale alitrunk. It is reputed to have been intercepted at ports in the British Isles. Another is *M. salomonis* (Linnaeus 1758), which is generally a much darker species with head and gaster black, alitrunk dark brown.

Messor capitatus (Latreille 1798)

Worker 4.0–12.0mm. *Queen* to 18mm. *Alates* September and October.

This large ant is black, the head sometimes with a reddish tinge, although tarsi and funiculus are often brown; the body has numerous short fine hairs. It is moderately shining, but the propodeum is slightly wrinkled and sculptured. The head is large, to massive; in the largest workers it is bigger than the whole of the rest of the body, a great broad nearly squarish globe that makes the ant look top-heavy. These

A *Messor capitatus* queen, one of the harvester ants.

are harvester ants, bringing back seeds into the nest and storing them in grain caches. This is nowhere near a native British species, but it and several other *Messor* species are a prominent part of the European ant fauna and could potentially be introduced into southern England with plant material brought back from holidays abroad or through the horticultural trade. Indeed, I was recently tweeted a picture of what is likely to be this ant 'dropped out of a supermarket vegetable' alive onto the carpet. It is also a popular species to be kept as 'pets' by ant enthusiasts, schools and colleges and is widely available from specialist suppliers, who will mail out a queen and 5–10 workers in the post. In France it has an Atlantic as well as a Mediterranean distribution, and occurs right up to Brittany – so it is conceivable that it could become established in southern or western Britain if given the chance. *Messor structor* (Latreille 1798) also occurs throughout much of France. It is hairier, and the base of the scape has a small sharp lobe-like outgrowth.

Myrmecina graminicola (**Latreille 1802**)

Worker 3.0–3.6mm. *Queen* 4.0–4.2mm. *Alates* April to August, but early spring individuals often remain in the nest for many weeks until suitable weather in summer.

A winged *Myrmecina graminicola* queen climbs a grass stem in readiness for flight.

This is a dark reddish-brown or sometimes almost black species; the mouthparts, antennae and stout legs are yellowish brown. The head, alitrunk and pedicel are heavily wrinkled and sculptured, while the gaster is smooth. The body is covered all over with slim yellow hairs. It is a sluggish slow-moving species that has a tendency to curl into a ball when disturbed, earning it the unlikely name of woodlouse ant. This has recently been suggested as an escape strategy – the ants have been observed curling into a ball and then rolling away downhill to evade spiders and other ants (Grasso *et al.* 2020). Small colonies numbering up to a few hundreds of workers and a few queens are made under stones and logs, often in company with other ant species – *Lasius alienus* and *L. flavus* most often. It scavenges and preys on small invertebrates, and although it does not attend aphids for honeydew, it has been recorded feeding on freshly cut apple slices (McClenaghan 2006). Its secrecy may have made it seem very scarce in the past, although it is also likely to be a genuine 'recent' arrival; Morley (1909) quotes Frederick Smith when reporting on its occurrences in the Isle of Wight: 'This is one of the rarest ants that occur in this country; the male was first discovered by Mr Curtis, in 1829, near Blackgang Chine, where it has subsequently been taken by other entomologists.' It is now known to be fairly common and widespread in southern England south of a line drawn from Colchester to Cheltenham, but is mainly coastal in south Wales. When I tried out a two-stroke blower-vac in my back garden I found several specimens of this secretive species (Jones 2004). A citizen-science pitfall-trapping project in Regent's Park in 2017 produced a very large number, including a few males and queens.

Myrmica hirsuta Elmes 1978

Worker about 4.0mm. *Queen* about 5.2mm. *Alates* August to October.

This species is very similar to *M. sabuleti*, inside nests of which it is an obligatory social parasite. Its postpetiole is larger and its body is more densely covered in slightly longer hairs. Workers are apparently extremely rare. In the British Isles it seems limited to a small area in Hampshire and Dorset, the type locality, from where the first specimens were described (Elmes 1978), and a single locality in west Wales.

Myrmica karavajevi (Arnoldi 1930) (formerly *Sifolinia laurae* Emery 1907)

Worker unknown. *Queen* 3.2–3.6mm. *Alates* July to September.

A small but distinct forwards-directed projection beneath the postpetiole, and a similar but smaller point on the petiole, should identify this species. The tibial spurs on middle and hind legs are simple or sometimes absent, whereas they are usually stout and pectinate in other *Myrmica* species. It is an obligate social parasite in the nests of *M. scabrinodis* and *M. sabuleti*, and is easily overlooked because the queens are slightly smaller than the host workers. In the British Isles it is very rare, restricted to the boggy lowland heaths of Dorset, Hampshire and Surrey, and the dry sandy Norfolk Brecklands, although there is also a record for Pembrokeshire.

Myrmica lobicornis Nylander 1846

Worker 4.0–5.0mm. *Queen* 5.0–5.5mm. *Alates* August and September.

This is a robust, rugose (roughly sculptured), dark reddish-brown species, sometimes nearly black; the head and gaster are sometimes darker than the alitrunk. It is named for a lobe-like outgrowth on the scape of the antenna, the size and shape of which vary across its geographic range. This is a widespread but local species, found in lowland heaths, rough grasslands and open woodlands throughout Britain, but not Ireland. In southern Europe it is regarded as a mountain species. Little is known of its feeding habits, but it is thought to mainly scavenge invertebrate material and is rarely found attending aphids for honeydew. It is also renowned for being reticent, retiring and non-aggressive, secretive even, so likely to be easily overlooked.

Myrmica lonae Finzi 1926

Worker 5.5–6.5mm. *Queen* 5.5–6.5mm. *Alates* August and September.

Previously, this was considered a variety of *M. sabuleti* with the lobe at the angle of the scape produced into an even more massive development. Its status in the British Isles is as yet unclear. It is apparently widespread in Europe, where it is often regarded as being less warmth-loving than *M. sabuleti*, and most Scandinavian specimens of *M. sabuleti* may in fact be attributable to this species (Collingwood 1979). However, recent work on mitochondrial DNA suggests this species is not distinguishable, and Ebsen *et al.* (2019) suggest it may need to be synonymised again.

Myrmica rubra (Linnaeus 1758)

Worker 4.5–5.5mm. *Queen* 4.5–7.0mm. *Alates* late July to September.

This ant is coloured reddish pink or yellowish brown, and it has faint sculpture. The legs and antennae are slightly paler reddish yellow. This is the default red ant in the British Isles, and usually readily distinguished by the slender and gently curving scape and by the smoother outline of the petiole. The propodeal spines are shorter than in other *Myrmica* species, and the area between them is more shining. It is common and widespread throughout Britain and Ireland, making small nests of usually a few hundred workers (but up to about 1,000 recorded) and a few queens, under logs and stones, small planks of wood and the like; Donisthorpe (1927a) notes a colony discovered in an old boot, and I have found colonies under a

Myrmica rubra worker on the prowl.

discarded work glove on the embankment of the Metropolitan Tube line, in a large butchered bone near Dagenham (east London), and inside the wrinkled pages of an old paperback book on a landfill site near Guildford (Surrey). Unlike so many of the more warmth-loving species it will nest in damp (but not wet) places such as gardens, woodlands, meadows and hedgerows. Workers visit aphid colonies after honeydew, and will also drink nectar direct from flowers. It is very active and aggressive and will readily sting, a sensation a bit like that from stinging nettles.

Myrmica ruginodis Nylander 1846

Worker 4.0–6.0mm. *Queen* 4.5–7.0mm. *Alates* July and August.

This ant comes in various shades of reddish brown, and, as its name suggests, the sculpture on the head, alitrunk and pedicel is longitudinally rugose. It is very similar to *M. rubra*, but has the petiole notched and the propodeal spines longer and sharper; the space between them has transverse furrows. This is possibly the commonest (or at least most widespread) ant in the British Isles, and it occurs in many habitats from parks and gardens to woods, moors and heaths from the south coast of England to the Scottish islands including the Outer Hebrides, St Kilda and Orkney. In Shetland it is the only ant species recorded. It occurs sporadically in Ireland, mostly around Galway and Waterford. In southern England it is particularly shade-tolerant. Nests comprising hundreds of workers (up to about 1,000 known) and 10–20 queens are made in logs, under stones, or in grass tussocks.

Myrmica sabuleti **Meinert 1861**

Worker 4.0–5.0mm. *Queen* 5.5–6.5mm. *Alates* August and September.

This is a robust, reddish-brown ant. The head and alitrunk are well sculptured, but the gaster is smooth and shining. Typically it is a species of chalk hills, sandy heaths and moorland slopes. It is widespread across all of England and Wales, but is more local in Scotland. It is sporadic, mostly coastal in Ireland. Nests of several hundred workers and 10–15 queens are made under logs and stones. Famously, it is the major host ant for the Large Blue butterfly, *Phengaris* (formerly *Maculinea*) *arion*.

Myrmica scabrinodis **Nylander 1846**

Worker 4.0–5.0mm. *Queen* 5.5–6.5mm. *Alates* July to September.

This ant is reddish yellow to dark brown, sometimes nearly black. The antennal scape is strongly right-angled, but has no large flange or outgrowth at the corner. This is a rather common ant throughout the British Isles, occurring in open grassy locations rather than in woods. Nests of a few hundred workers and a few queens are made in the soil, or under rocks and stones. It preys on small invertebrates, including the brood of other ants, notably *Lasius flavus*. Recent work on mitochondrial DNA suggests that in Europe at least there are two cryptic species under this name; in Denmark these may be split along habitat lines, with one preferring dry areas, the other tolerating sphagnum bogs and lake edges, although both types occur together in the Netherlands (Ebsen *et al.* 2019).

Myrmica schencki **Viereck 1903**

Worker 4.0–5.5mm. *Queen* 5.0–6.0mm. *Alates* July to September.

This species is reddish brown, with the head and gaster sometimes darker. The large blunt tooth on the bend of the antennal scape is accommodated by a larger than usual bend in the frontal carina, and a rounder deeper hole in the front of the head. This gives the impression of a narrower forehead between the antennae, the frons appearing to be about one-quarter the width of the head. It is a relatively scarce ant of southern Britain, where it makes small nests of up to a few hundred workers and a single queen in sandy or rocky places, usually in warm sunny positions. Most records are from the Thames estuary and Thames basin, but outliers extend to Norfolk, Peterborough, Cheltenham and Swansea. It also occurs in western Ireland. It is possibly a nocturnal predator of other ants (Allen 2009).

Myrmica specioides **Bondroit 1918 (formerly called**
M. puerilis **Strärke 1942). There is some controversy on the**
name of this species, with some myrmecologists using
the name ***M. bessarabica*** **Nasonov 1889.**

Worker 3.0–4.5mm. *Queen* 5.0–5.5mm. *Alates* August and September.

This is a small but aggressive reddish-yellow species. It is likely a recent colonist, first found in England, at Deal in east Kent, in 1961 (Collingwood 1962) but with older museum specimens later discovered to have been taken in Whitstable in 1958 (Allen 2009). It is widespread in Europe, including all of France. It is still very restricted here, but is now recorded around the coast from Great Yarmouth (Norfolk) to the Isle of Wight, with most records in the Thames estuary. It also occurs more inland in Kent, including parks and gardens. I found it in a farmer's bulldozed sand heap at Manston, near Ramsgate, on 25 May 2008 in a most unlikely site surrounded by extensive arable fields and industrial estates. It makes its nests in sandy or gravel banks, with up to 2,500 workers recorded and several queens. The majority of its food seems to be in the form of workers and brood of other ants, notably *Lasius* species. It is reputed to sting given half the chance.

Myrmica sulcinodis Nylander 1846

Worker 4.0–6.0mm. *Queen* 5.5–6.8mm. *Alates* July and August.

This ant is deep reddish, often with the gaster darker, sometimes brownish to black. It is a particularly strongly sculptured species, with longitudinal wrinkles along the head and alitrunk, and with the frontal triangle (usually smooth and shining in *Myrmica* species) with at least a few wrinkles along its length. This is a rather scarce species of upland moors and lowland bogs. It has a rather disjunct distribution in Britain, occurring sparingly through Surrey, Hampshire and Dorset down into the West Country, then through Wales, northern England and up into Scotland (but not Ireland). It makes small nests of a few hundred workers and one queen under stones in clumps of peat. In the rest of Europe it is regarded as a mountain species. It is probably declining in much of its southern English range; many former Surrey heathland sites have now become unsuitable (Pontin 2005).

Myrmica vandeli Bondroit 1920

Worker 4.0–5.5mm. *Queen* 5.5–6.5mm. *Alates* August and September.

This ant is hairier than other *Myrmica* species, the petiole with 10–20 (or more) long, thin, curved hairs. The petiole and postpetiole show perhaps less sculpture. The spurs on the middle and hind legs are reduced, and lacking the pectination (comb-like arrangement of fine teeth) found in other species. It is extremely similar to *M. scabrinodis*, in nests of which it often coexists, suggesting that it is a temporary social parasite. It was only recently determined as a British species. Workers cannot easily be separated, but the males have longer antennal scapes, like those of *M. sabuleti* (Elmes *et al.* 2003). Little is known about its biology, but it seems to favour warm and humid sites such as damp meadows, marshes and heathland bogs. In Europe it is regarded as an upland species, nesting in grass tussocks. So far there are records only from southern England and south Wales.

Solenopsis fugax (Latreille 1798), sometimes called fugitive ant

Worker 1.5–3.0mm. *Queen* 6.0–6.5mm. *Alates* August to October.

This is a tiny yellowish or pale brown ant, though the queens are darker. The head and alitrunk are shining, but distinctly punctured. The 10-segmented antennae are a key identifying feature. This shy and retiring ant is seldom found above ground, but it is a fierce predator of other ants' brood and is often found in company with larger *Formica* and *Lasius* species. The 'escape' tunnels into the *Solenopsis* nest are very narrow, and too small for other ants to get through. Most British colonies are coastal, and this species makes large, populous nests with 100,000 workers and numerous queens, in

Worker of *Solenopsis fugax* stealing brood from a *Tapinoma* nest.

Soldier and minor workers of *Pheidole megacephala*, showing the huge size disparity.

dense clay or rock cliffs. It seems to be very rare, and has only been found in four localities since 1969 – Isle of Wight, Weston-super-Mare (Somerset), Salcombe (Devon) and Hartlip (east Kent). It is right on the edge of its European range here. Its subterranean nature probably makes it easy to overlook. Another species, *S. monticola* Bernard 1950, is sometimes mentioned as occurring in the Channel Islands, but this is now regarded as a synonym of *S. fugax*.

Pheidole megacephala (**Fabricius 1793**)

Worker 2.5–4.0mm, but of two distinct sizes in the colony.
Queen 6.5–9.0mm. *Alates* June and July in outdoor colonies around the Mediterranean.

The largest workers have a massive broad head, giving it the name big-headed ant in North America. It is brown to brownish black, relatively shining, but with scattered long bristles over the body. Large workers are soldiers, and as well as the massive heart-shaped head, the thoracic alitrunk is also enlarged. It was originally described from Mauritius, and this is probably an African species, but it has been accidentally transported all around the world and is a major invasive pest in North America, Australia and elsewhere. In the British Isles it can only survive in heated buildings, and most records are of temporarily established colonies in hothouses.

Stenamma debile (**Förster 1850**)

Worker 3.5–4.0mm. *Queen* 4.2–4.8mm.
Alates August to October.

This is a small, narrow, elegant ant, red to dusky brown, with slim petiole and minute eyes. It is extremely similar to *S. westwoodii* (below), and the distinctions were only made apparent after Dubois (1993, 1998) published redescriptions. Identifications are sometimes best confirmed using males – mandibles of male *S. debile* are slightly cylindrical and have three teeth, those of *S. westwoodii* are triangular and have five. This is a secretive, mostly subterranean species, making small nests of up to about 100 workers under stones or small logs, or in leaf litter, mostly in well-drained but shady woods and hedgerows. It feigns death when attacked by other ants. It is fairly widespread in southern and eastern England, but outside of the Thames basin it seems to be mostly coastal around East Anglia, the West Country and Wales. Almost all previous records of *S. westwoodii* actually refer to this species – all mine do.

Museum specimen of *Stenamma debile*, showing narrow stalk-like petiole.

Stenamma westwoodii **Westwood 1840**

This species is virtually identical to *S. debile*, with which it was previously confused. It is reportedly more 'leggy' than *S. debile*, in that it has slightly longer legs and antennae, but this is highly comparative and difficult to appreciate unless specimens of both species are examined together. It seems to be much rarer than *S. debile*, with confirmed records from a handful of sites across southern England from Suffolk to Somerset. It is regarded as extremely scarce in Europe, and is known only from Belgium and the Netherlands. Its biology is probably very similar to *S. debile*. Incidentally, it might appear that this species was named by John Obadiah Westwood (1805–1893) after himself, something which would then have been regarded (and it still is) as being jolly bad form. In fact he did not. The name was first used by his friend, colleague and founder member of the Entomological Society of London, James Francis Stephens (1792–1852). It was officially published on page 356 of the *Catalogue of British Insects* (Stephens 1829) as genus number 666 (number 92 in the Hymenoptera) *Stenamma*, and species number 4,838 *westwoodii*, where

Stephens indicates the authority is 'mihi', from the Latin meaning 'to me', and indicating 'my work' or 'I did this'. In other words he named it so. He also indicated that the specimens for which he intended the name were in the collection of Dr Westwood. But, crucially, he did not describe the species – so it didn't count. Today, the authority and date quoted after a scientific name refer to the first *published description* of the organism. As it turns out, the first time this name was attached to a description was in 1840 when Westwood himself published the second volume of his magnum opus *An Introduction to the Modern Classification of Insects*. Here, on page 219, he describes the mandible and palp structure, on page 226 he gives the relevant figures, and on page 83 of the supplementary catalogue published at the back of the volume he gives the first brief description, including the important point that the first joint of the peduncle (as he calls the pedicel) is elongate. True, he acknowledges that the genus name *Stenamma* is his own coining, but he is clear in showing that the epithet *westwoodii* is from Stephens' *Catalogue*. Even so, Westwood still appears as the describing authority for this species, as justified by the International Code of Zoological Nomenclature.

Strongylognathus testaceus (**Schenck 1852**)

Worker 2.0–3.6mm. *Queen* 3.5–3.8mm. *Alates* July and August.

This small yellowish ant has a long rectangular head with pronounced rear corners, and the back of the head is fairly deeply indented. The most distinctive feature of this species is seen in the

Strongylognathus testaceus, showing the narrow curved jaws and pronounced rear corners of the head.

'falcate' mandibles, which are narrow, curved and sharply pointed, very different to the broad triangular multi-toothed jaws of most other ants. *Strongylognathus* is an obligate social parasite in the nests of *Tetramorium caespitum*, where they stand out strongly, being pale compared with their dark hosts. The host queen is killed and the nest's workers are usurped into caring for the *Strongylognathus* brood. It is very rare – known only from a few sites in Hampshire, Dorset and Devon.

Temnothorax nylanderi (Förster 1850)

Worker 2.3–3.4mm. *Queen* 4.2–4.7mm. *Alates* July and August.

This ant is small, delicate, yellow to pale brown, with the head sometimes darker, and the gaster usually with the first segment pale at the base (often appearing as two pale spots), but with a darkened band across the centre. The alitrunk is indented across the middle, a shallow suture separating the mesonotum and propodeum, and visible in side view as a depression. It is distinct from others in the genus by having the antennal segments unicolorous, the club not being darkened. This is a local species in southern England, East Anglia and south Wales. It makes nests of 100–200 workers and a single queen in small cavities in rotten logs, under bark, or in the abandoned tunnels of wood-boring beetles. There are reports of it nesting in hollow acorns, old pine cones and snail shells, and I have found it in the hollowed mouldering spaces between the concentric layers of the hard, dry, Cramp-ball fungus *Daldinia concentrica* (Jones 2016).

Lone worker of
Temnothorax nylanderi.

Temnothorax interruptus (**Schenck 1852**)

Worker 2.3–3.4mm. *Queen* 3.7–4.2mm. *Alates* July and August.

This species is pale yellowish, with vague darker smudges on the first gastral segment – the interrupted band of its name. The antennal club is distinctly darker than the funiculus. The propodeal spines are long and rather curved. This rare ant makes small nests of a few hundred workers in the soil. It is very restricted in southern England, usually near the coast. It has recently been found only at Dungeness, the Dorset heaths and the New Forest.

Temnothorax unifasciatus (**Latreille 1798**)

Worker 2.8–3.5mm. *Queen* 4.0–4.5mm. *Alates* July and August.

This yellowish ant sometimes has the head vaguely darker, and the gaster has a distinctive dark band across the end of the first segment, giving it a brightly marked appearance. The propodeal spines are rather short and stout. The alitrunk is flatter than in *T. interruptus*. It is well known from the Channel Islands, and a small colony has been recorded in a Chiswick (west London) garden (Fox 2011), but this may have been a casual import. It is widespread in France and is a potential future colonist. It makes small nests of up to about 200 workers and one queen in warm, dry, open sites, under rocks and stones, in rotten logs and tree stumps.

Temnothorax albipennis (**Curtis 1854**) (**formerly misidentified as** *T. tuberum* **by British entomologists**)

Worker 2.3–3.4mm. *Queen* 3.7–4.5mm. *Alates* July and August.

This is a pale yellowish ant, dirty brown in some specimens, with the head often darker. The propodeal spines are short and stout. The darkening on the first gastral segment is poorly defined. This uncommon species makes small nests of up to about 350 workers and one queen under rocks and stones, or in cracks in the ground, snail shells or hollow stems. Almost all British records are from coastal sites between Deal in Kent and Llandudno in north Wales. The shingle of Dungeness appears to be a particular stronghold, where it nests in dead gorse and broom stems (Allen 2009).

Temnothorax parvulus (**Schenck 1852**) **and** *T. affinis* (**Mayr 1855**)

Both of these species are widespread in France and are potential future colonists. *T. parvulus* resembles a slightly less strongly marked

T. nylanderi. T. affinis has rather long and slender propodeal spines. More details can be found in Blatrix *et al.* (2013).

Tetramorium caespitum (**Linnaeus 1758**)

Worker 2.5–4.0mm. *Queen* 5.5–8.0mm. *Alates* June to October.

This is a small, dark brown, sometimes almost black, robust and rather aggressive ant. Although small, it can open its jaws wide enough to bite human skin. The propodeal spines are short and very stout. The legs and antennae are mid to light brown. The angular front corners of the thorax are very distinctive. This is an uncommon warmth-loving species, and apart from the lowland heaths of Surrey, Hampshire and Dorset, and the sandy Brecklands of East Anglia, most sites are coastal. It is recorded from Norfolk, around the south coast of England, the coast of Wales to Anglesey, Isle of Man, and with outlier localities in Scotland. The largest nests may contain 10,000 workers (but usually only one queen) under stones and logs or in rock crevices, in areas of sparse vegetation, and often in loose crumbling soil. It is regarded as being a good indicator of high-quality heathland (Pontin 2005), and can survive heathland fires and other disturbance. The last time I saw this species was under old wooden sleepers, demolished building rubble and discarded wooden planks on a military shooting range near Longmoor, Hampshire, where regular vehicular and foot traffic kept the heathland free of scrub and created plenty of areas of sparse vegetation. This is

Worker of *Tetramorium caespitum*, showing the broad angular shoulders.

the only native British ant that collects and stores seeds (heather, grasses, birch etc.), and feeds them directly to its larvae. It also scavenges, captures invertebrate prey and collects honeydew. There are now thought to be six cryptic species allied to *T. caespitum* in the Western Palaearctic. The most likely other British species is *T. impurum* (Förster 1850), and indeed this species has recently been reported from Guernsey (Attewell & Wagner 2019). Distinctions are subtle, to say the least, based on a combination of molecular and morphological measures. The British fauna awaits an unravelling of this complexity before any further identification key can be created.

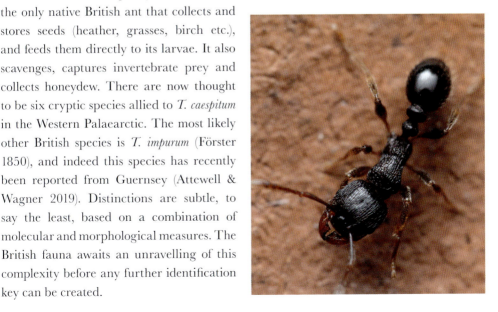

Tetramorium bicarinatum (**Nylander 1846**), *T. caldarium* **Roger 1857, and** *T. simillimum* **Smith 1851**

Tetramorium is a huge genus worldwide, and several species have been accidentally transported around the world; some have become temporarily established in heated buildings in northern Europe. These three have reputedly been found at various times in the British Isles. *T. bicarinatum* is reddish yellow with contrasting dark gaster; the frontal carinae are extended backwards across the top of the head nearly to the occiput. *T. simillimum* is similar but with the body more finely rugose. *T. caldarium* is all over reddish brown. Others may well occur here in the future.

Tetramorium (**formerly** *Anergates*) *atratulum* (**Schenck 1852**)

There are no workers in this species. *Queen* 2.3–2.5mm. *Alates* May to September.

This species is dark brown with yellowish legs. The jaws, which are narrower than the usual triangular myrmicine form and lack the serrated inner edge, appear narrowly leaf-shaped – this is typical of social parasites. It occurs only in the nests of *T. caespitum*. The means of colony usurpation is unclear. A queen *T. atratulum* (or several sisters) may take over a queenless nest of *T. caespitum*, or their arrival may lead to the host queen being killed or starved. *T. atratulum* queens are smaller than their host workers, but soon become physogastric, large and swollen with eggs. The larvae are more hairy, slightly narrowed at the waste, and yellowish, contrasting with any remaining host brood, which is smooth, barrel-shaped and white. The male is wingless, 2.2–2.3mm, and pale dull grey. The males are sluggish, pupoidal, and mate with new females inside the nest. This is a very scarce species, with only a few recorded localities on the Surrey heaths, New Forest, Purbeck and Dungeness.

Strumigenys perplexa (**Smith 1876**)

Worker 1.5–2.0mm.

New ant species are constantly being added to the British list. In 2020 Andy Marquis found several colonies of a species of trap-jaw ant living out of doors in Guernsey, feeding on springtails. They were originally thought to be *Strumigenys silvestrii* (Emery 1906), a South American species which has been accidentally transported to North America, where it is now fairly widespread, and which has recently been found on Madeira and in mainland Portugal. However, they

Strumigenys perplexa, recently discovered in Guernsey.

have now been identified as *S. perplexa*, an Australasian species likely to have arrived in Europe with imported plants (Hamer *et al.* 2021). It is 1.5–2.0mm long, pale yellowish brown, covered with short stout scale-like hairs, and is immediately identifiable as different from all our other British ants by its extra-long heart-shaped head and very long, narrow jaws which are abruptly curved in typical trap-jaw style at the tips. Whether this species will spread to mainland Europe or Britain remains to be seen.

Cardiocondyla britteni: the type specimen (the original individual specimen from which the species was described).

Cardiocondyla britteni Crawley 1920

Worker 1.8mm.

This is a tiny shining red-brown ant, with scape and legs yellowish, head and gaster darker than alitrunk. The funicular segments are short and stout, the final segment of the antennae being large and broad, as long as the eight previous segments together. The petiole is narrow, the postpetiole globular. This species cannot, by any stretch of the imagination, be properly considered a British ant. *Cardiocondyla* is a large worldwide genus of about 70 species, mostly in the tropics of Asia and Africa. None is native to northern Europe, let alone the British Isles. However, by a strange quirk of fate, this species has only ever been recorded from England. Therefore it is, sort of, a British ant. It's a bit of a conundrum. The only specimen ever discovered, the original 'type' from which the species was formally described, was found by Mr H. Britten in some butter beans in West Didsbury, Manchester, on 12 May 1919. Even at the time Crawley (1920), who described it new to science, knew it was an imported species, but other than guessing it had

come from tropical America where the beans were native, he could make no other geographical observation. It is likely to be a variant of some species in the *C. minutior* group, an Asian not an American phylogeny. Butter beans are now grown in Africa and India, but where the Didsbury supply came from is still rather guesswork. The ant's true status still needs to be addressed. The single type specimen remains in the Oxford University Museum of Natural History, should anyone care to go and have a look at it.

Subfamily Formicinae

The Formicinae represent perhaps the most important subfamily of ants in the British Isles, since the most numerous and most obvious species here (*Lasius* and *Formica*) make the largest nests. All members of the subfamily have a single pedicel node, and the gaster ends with a distinct, round, hair-ringed orifice – the acidopore – through which formic acid is squirted.

Plagiolepis taurica (**Santschi 1920**) (**formerly called** *P. vindobonensis* **Lomnicki 1925**)

Worker 1.0–2.0mm. *Queen* 3.0–4.0mm. *Alates* June and July.

This is a minute, pale or dark brown species, its body shining, with sparse pubescence. Small isolated nests containing a few hundred workers and several queens are made in warm, dry, open areas, under stones and rocks, or in cracks on coastal cliffs. It is known from the Channel Islands, and is widespread in France and Belgium, but has not yet been found on mainland Britain.

Plagiolepis alluaudi **Emery 1894**

Worker 1.5mm. *Queen* 3.0mm.

The gaster of this pale yellow ant becomes translucent after feeding. The head is almost circular when viewed from directly in front. This is a tropical species, most likely originally from Madagascar, that has been transported around the world. It has become a pest on Pacific islands, and in European greenhouses. It was found at the Stratford-upon-Avon butterfly house, and also at London Zoo, in 2015 and 2016, and in Threave Garden in south-west Scotland, in 2019, presumably introduced on horticultural material and able to establish colonies in the heated glasshouses (Mallumphy 2016, Hancock & Robinson 2021).

Polyergus rufescens **Latreille 1804**
Worker 5.0–7.0mm. *Queen* 7.0–9.0mm. *Alates* July and August.

This characteristic species is entirely reddish brown. The propodeum is domed, the petiole large, thick and blunt; the petiole and gaster have numerous outstanding bristles. The most distinctive feature is the narrow sabre-like mandibles. A 'slave-maker', it raids *Formica* nests and makes off with the brood, most usually pupae; when the *Formica* workers emerge they take on foraging, brood care and defence duties in the now mixed-species nest. This species has not yet been recorded in Britain, but it is widespread in Europe, including France and Belgium, and two of its recorded host species are *Formica fusca* and *F. cunicularia*, which are both common in southern England. A potential colonist in the future perhaps?

Paratrechina longicornis **Latreille 1802**
Worker 2.5–3.5mm. *Queen* 5.0–5.5mm.

This is a blackish-brown, rather long and slender ant, covered in sparse but prominent pale bristles. The head is narrow in proportion to the alitrunk, which is rather parallel-sided. The antennae and legs are very long and slim. This is a pan-tropical species carried around the world by human commerce, possibly originally native to south Asia. It is seemingly unable to establish colonies out of doors in northern Europe, but colonies are occasionally established in heated buildings or hothouses.

Nylanderia (**formerly** *Paratrechina*) *vividula* (**Nylander 1846**)
Worker 2.0–3.0mm. *Queen* 3.5–5.5mm.

This usually dirty yellowish-brown ant can vary to dull black; the head and gaster are generally darker than the alitrunk. The body is covered in short but crowded pale bristles. This is another tramp species that has been spread around the world by human commerce; it is thought to be originally native to the Americas. It is unable to establish outdoor colonies in the British Isles, but can survive in heated buildings and hothouses.

Formica aquilonia **Yarrow 1955**
Worker 4.0–8.5mm. *Queen* 8.0–10.0mm. *Alates* May to July.

The head and alitrunk are bicoloured, reddish orange-brown (bright in life, but perhaps not as much as *F. rufa*), marked with slightly darker blotches; the gaster is dark brown. The legs are dark reddish brown, the antennae brownish black. Distinctions from *F. lugubris*, with which it overlaps geographically, are frustratingly subtle. It is found only in the Caledonian pine forests and undisturbed birch groves of the Scottish Highlands, and in County Armagh in Northern Ireland. In the rest of western Europe, this ant occurs in Denmark and Fennoscandia into Siberia, where it inhabits open coniferous woodland, but elsewhere in Europe it is considered an alpine insect. Large nests of heaped leaf litter (mostly pine needles) are made on well-drained sites or woodland slopes, and although they usually have some exposure to sun, *F. aquilonia* appears to be more shade-tolerant than other wood ants. Long trails lead up the trees to sap-sucking Homoptera, and to other nests which form one large colony. They also scavenge insect remains from the forest floor.

Formica cunicularia Latreille 1798

Worker 4.0–6.5mm. *Queen* 7.5–9.0mm. *Alates* June to August.

This active ant is dull greyish brown, but with some reddish-brown markings on the sides of the alitrunk and at least the underside of the head; rarely these are entirely red. The legs are variable reddish brown, and the antennae are often slightly darker. The petiole scale lacks any upstanding hairs on its upper margin. This is a widespread species in southern England, but away from the south-east and Thames basin it is more or less coastal, with records from Norfolk around to the Lleyn Peninsula in north Wales. It prefers open, sunny downland or heathland sites. Small nests of 1,000–1,800 workers with

Formica cunicularia, an active red and brown species.

a single queen are made under rocks and logs, or small earth mounds are erected in the grass root-thatch. This is a frequent brownfield ant in the London area. I first found it nesting inside small stop-cock water valve inspection chambers on a derelict covered reservoir in Forest Hill, south-east London, and it is quite at home nesting in bulldozed rubble heaps on abandoned industrial sites. Workers forage on flowers after nectar, visit aphid colonies, and scavenge small invertebrate prey.

Formica exsecta Nylander 1846, the narrow-headed ant
Worker 4.5–7.5mm. *Queen* 7.5–9.5mm. *Alates* July.

The head and alitrunk are brown with varying amounts of reddish blotching; the gaster is dark brown. The deeply excised rear margin to the head is very distinctive. Viewed from the front, the petiole scale is notched on top. The eyes are hairy. This is a very scarce species with a highly disjunct distribution in Britain (but it does not occur in Ireland). During the 20th century it occurred, sporadically, in the lowland heaths of Dorset, Hampshire and the Isle of Wight, but recent surveys have failed to rediscover it there. One specimen from Woking in Surrey, in 1913, possibly suggests it used to occur in that county too (Pontin 2005). There is now, in England, one known colony – in Devon. It also occurs very rarely in the Scottish Highlands, with most sites known from Abernethy, Glenmore and Rothiemurchus, also Deeside and Rannoch Moor. Colony formation can be by budding, or by temporary social parasitism in nests of

Formica exsecta is known from only one site in England, although it is hanging on a bit better in Scotland.

F. fusca or *F. lemani*. It makes small mound-nests, about 30cm across, in open heathland, containing up to about 1,000 workers and several queens.

Formica fusca Linnaeus 1758

Worker 4.5–7.0mm. *Queen* 7.0–9.5mm. *Alates* June to August.

This is a dark greyish- or brownish-black ant, with antennae, tarsi and knees slightly reddish. It is a common and widespread species throughout England and Wales. In Scotland and northern England it is usually replaced by *F. lemani*. The main distinction is in the lack of small but stout upstanding hairs on the back ridge of the head and on the anterior part of the alitrunk. This distinction from *F. lemani* is difficult, and most authors suggest a minimum of five specimens should be microscopically examined to be sure that they are not *F. lemani* that have had their hairs abraded. The sculpture of the frons is very fine, and in the area between the ocelli, lateral carinae and the eyes, individual punctures are difficult to see, even under high magnification. The nests, usually under logs or stones, or in tree stumps, are made in open woodland, verges, heathland and moors, and comprise a single queen and a few hundred, up to 1,000, workers. In England, at least, this is the default large black ant of open woodlands and rough grassy places. It runs rapidly, but is not at all aggressive, and it retreats timidly as soon as a nest is uncovered. Workers collect aphid honeydew, but also scavenge or prey on small invertebrates.

Formica lemani Bondroit 1917

Worker 4.5–7.0mm. *Queen* 7.0–9.5mm. *Alates* July and August.

This large black ant is virtually identical to *F. fusca*, but can be distinguished by the presence of numerous short stubby hairs on the occiput of the head and on the promesonotal dorsum (front half of the alitrunk). These are small, but pronounced, and a seemingly hard-and-fast specific character. The sculpture on the frons is coarser than in *F. fusca*, and individual punctures are readily seen, the interspaces between them appearing more shining. This is a northern and western species in the British Isles, occurring in Ireland, Scotland and northern England (where it is often the commonest ant species), and broadly overlapping with *F. fusca* in Wales and the West Country. As with *F. fusca*, small nests of up to about 1,000 workers and one queen are made in open woodlands, heaths and moors.

Formica lugubris Zetterstedt 1840

Worker 4.5–9.0mm. *Queen* 9.5–10.5mm. *Alates* May to July.

The head is dark brown or black, but with the underside broadly reddish brown. The alitrunk is reddish with a large irregular dark patch on the promesonotum. The petiole is reddish brown, and the gaster, legs and antennae are dark brown. This is a rather scarce upland ant, occurring in north Wales, northern England, Scotland, and also in southern Ireland. It is a species of conifer and broadleaf woodland, but also nests well away from the treeline, sometimes in very wet sphagnum bog. Colony foundation is sometimes by nest-budding, but also by temporary social parasitism. In such cases a female *F. lugubris* is adopted by a *F. lemani* nest, and after killing the *F. lemani* queen she becomes the exclusive egg-layer. Eventually the host workers are lost by attrition and the colony gradually becomes wholly *F. lugubris*. As well as being able to survive in much wetter regions, *F. lugubris* is able to forage at much lower temperatures than *F. rufa*, and replaces it across much of Scandinavia and northern Eurasia. In north-western Spain, this species was a prime disperser of seeds of the Stinking Hellebore *Helleborus foetidus*. It was deliberately introduced into Quebec forests in the 1970s to control defoliating insects (Finnegan 1975), and nests were still active a decade later (Jurgensen *et al.* 2005). An interesting survey of this species in Cumbria is given by Robinson & Woodgate (2004), who report a colony density of 5.7 nests per hectare and some giant nests up to 9m in diameter and nearly 3m high on rocky outcrops.

Formica picea Nylander 1846, sometimes known as shining bog ant (formerly called *F. candida* Smith 1878, and *F. transkaucasica* Nasonov 1889)

Worker 4.5–6.0mm. *Queen* 8.0–9.0mm. *Alates* July and August.

This is a dark, slender species, black or very dark brown, more or less shining, with pubescence on the gaster very diffuse. The frontal triangle is shining. The antennae are shorter than in other species, especially the second funicular segment, which is not much longer than its maximum width, and shorter than the third. It is a very scarce species of lowland bogs in England and Wales, and is known mainly from Dorset and Hampshire (mostly around Wareham and the New Forest), with a few colonies in south Wales. There was a well-established but isolated group of nests on Colony Bog, in Surrey (Denton & Collins 2004); this is a military area, so access to survey the site is difficult. Contrary to the usual notion that ants are warmth-loving creatures of well-drained soil, this is perhaps the most hygrophilous ant species in the British Isles, nesting in treacherous sphagnum quagmires which are difficult to survey. Nests are usually constructed within and around a tussock of Purple Moor-grass or among moss. Colonies are thought to be mostly small, with up to about 1,000 workers and one or a very few queens.

Formica pratensis Retzius 1783

Worker 4.5–9.5mm. *Queen* 9.5–11.3mm. *Alates* May to August.

The head is red with the frons and occiput matt black or dark brown. The petiole and alitrunk are reddish with a variable dark blotch on the promesonotum. The gaster is blackish brown, and fairly thickly pubescent. The legs and antennae are dark. This species is now probably extinct on the British mainland. It occurred in the Wareham and Bournemouth areas in Dorset, but was last located there in 1987. It occurs on the Channel Islands and is widespread in France. Colony foundation is usually by temporary social parasitism where a *F. pratensis* queen enters a *F. cunicularia* or *F. fusca* nest and takes over egg-laying after the host queen is eliminated.

Formica rufa Linnaeus 1761

Worker 4.5–9.0mm. *Queen* 9.5–10.5mm. *Alates* April and May.

This is a strongly marked species – fairly bright red and black or very dark brown. The face and jowls are red with the frons and occiput dark. The alitrunk and petiole are red with a variable dark mark on the promesonotum. The gaster is dark. Legs and antennae are dark, but with variable lighter areas. This is the 'common' wood ant of lowland Britain, occurring mostly in southern England and Wales, though with scattered records as far as Cumbria, but not in Ireland. It does not appear to have been recorded from Scotland. Huge mound-nests reaching more than 1m high and 2–3m in diameter may contain 400,000 workers bringing 60,000 food items a day back to the nest. In Britain most nests contain more than one queen. Sometimes nests are interconnected by ant trail lines, forming one large colony. Most colonies are in mixed or conifer woodlands, extending out to heathy scrub nearby. The large nest mounds are usually thatched with fallen pine needles, as well as small twigs and leaf fragments. Despite being the nominate wood ant in the British Isles, it may not be as common as it once was, and there are indications that its range is contracting.

In a survey of the species in the Lake District, Robinson (2001) reported that it had virtually disappeared since the 1950s, partly blaming abnormally heavy May rainfall in 1966–1968 for many local nest extinctions. Changes in lowland woodland management may mean that canopy shading is making sites cooler and damper, and less attractive to the ants. Accidental and deliberate nest destruction are regularly reported, and agricultural and urban encroachment have been blamed for habitat fragmentation. A large nest in bright sunshine in late spring or early summer is an awe-inspiring sight, and the rustling sounds of the ants, mustered together in the warm sun, or their footsteps from their supply trails through the dead leaves of the woodland floor are, for me, an evocation of a blissful childhood spent exploring the Sussex Weald during the 1960s and 1970s.

Formica rufa workers, showing the sometimes bright red markings on alitrunk and face.

Formica rufibarbis Fabricius 1793

Worker 4.5–7.0mm. *Queen* 8.5–10.5mm. *Alates* June and July.

This species is bicoloured, red and brownish black. The top and rear of the head, a mark on the promesonotum, and the gaster are dark. The legs and antennae are variably light reddish (especially in life) or dark brownish (in museum specimens). The short stout hairs on the top of the alitrunk are distinctive under the microscope. The petiole scale also has some bristles along its upper margin. This is one of Britain's rarest ants, on the verge of extinction, with colonies known only from

Surrey and Scilly. Numerous historical Surrey sites now appear to be unsuitable for the ant, which is currently recorded only from three or four colonies on the hot dry sandy heaths of Chobham Common, and the Stickledown Rifle Range near Camberley (Pontin 2005). Allen (2009) makes a cautionary observation about the Kent records for this species, suggesting that specimens which key out as *F. rufibarbis* may simply be a few aberrant rather red and hairy examples of *F. cunicularia*, especially as only singletons are ever found, and usually in areas where *F. cunicularia* is common. Small nests containing up to about 500 workers and two or three queens are usually made in the ground, or under stones.

Formica sanguinea **Latreille 1798**

Worker 6.0–9.0mm. *Queen* 9.0–11.0mm. *Alates* July.

As its name suggests, parts of this ant are bright red, notably the front of the head, alitrunk, petiole and legs. These contrast with the slate-grey gaster, although they fade slightly to a reddish brown in museum specimens. Southern European races appear to be more brightly reddish over more of their bodies, with specimens from Arctic regions being predominantly dark, with nearly all-black head and dusky alitrunk. In Britain this species has a highly disjunct distribution, with a southern population across the Weald and heaths of Sussex, Surrey, Wiltshire, Hampshire and Dorset, and Scottish colonies in the central Highlands. It does not occur in Ireland. There are also some outlier sites on the Welsh borders. It was previously known from Kent, but not discovered in recent keen survey activity (Allen 2009). Open mixed or conifer woods, or heaths, are preferred. Rather than making large thatched nest domes, small nests are made under logs, in tree stumps or in small piles of leaf litter. Nests are usually made in direct sunlight, and the colony will move or die out if the nest becomes shaded. In Britain this ant is unique for being a slave-maker, raiding nests of *F. fusca* or *F. lemani*, and taking larvae and pupae back to the home nest either as food or to rear as auxiliary workers. It has a reputation for being a very aggressive species, biting strongly rather than spraying formic acid.

Lasius alienus (**Förster 1850**)

Worker 3.0–4.2mm. *Queen* 8.0–9.0mm. *Alates* July and August.

This ant is black, although sometimes dirty greyish yellow through brown. The body is moderately covered with pubescence. The scapes and front tibiae lack the erect hairs so prominent in *L. niger* and *L. platythorax*. Separation from *L. psammophilus* (see below) is tricky, but *L. alienus* appears darker in life. This is a moderately widespread species in central and southern England and Wales, with a few records from Scotland and Ireland. It seemingly prefers warm dry habitats like chalk. It feeds on the honeydew of underground and arboreal aphids.

Lasius psammophilus Seifert 1992

Worker 3.0–4.2mm. *Queen* 8.0–9.0mm. *Alates* July and August.

This species is extremely similar to *L. alienus* (above), but appears slightly paler in life, and with a few short erect hairs on the rear slope of the propodeum. It is recorded in England and Wales, usually on more sandy or gravelly substrates than *L. alienus*, with many records being coastal, or in the band of sandy lowland heaths from the Brecklands through Surrey and Hampshire. Confusion between the two species probably means it is under-recorded. Its large underground nests may contain up to 12,000 workers.

Lasius brunneus (Latreille 1798)

Worker 3.2–4.5mm. *Queen* 8.0–9.0mm. *Alates* May to July.

This small secretive ant is bicoloured, the dark brown gaster contrasting with the reddish-brown alitrunk and head. It is a rather local species which nests in the interior of old trees, under loose bark, or in stumps and logs. It is timid and non-aggressive and disappears quickly if its galleries are uncovered when peeling bark off rotten trees. It was previously considered a rare ant of the old woodlands in Surrey, Berkshire and other central Home Counties, and the Severn Vale, but it appears to have spread widely during the last 50 years, expanding its range north along the Welsh Marches, north-east into Essex and East Anglia (Alexander 2006), and south into Wiltshire, Hampshire and Sussex (Alexander & Taylor 1998). It circumnavigated London to arrive in west Kent some time in the late 1980s (Jones 1996), and is now common in woodlands in south-east London. It is now also recorded from Somerset (Lush 2007), Wiltshire (Blacker & Collingwood 2002), east Kent and south Hampshire (Denton 2018). I have found this species under the bark of old London Plane *Platanus* × *hispanica* trees in highly urban central London at Elephant and Castle (in 2013), and under Sycamore *Acer pseudoplatanus* bark at Olympia (in 2010 and 2011). In May 2003 I was given some specimens infesting a biscuit tin in a house in Guildford, Surrey (Jones 2003).

Lasius emarginatus (**Olivier 1791**)

Worker 2.5–5.0mm. *Queen* 7.0–8.0mm. *Alates* July and August.

This ant is superficially similar to *L. brunneus* in being bicoloured, but the alitrunk is a much brighter red. The antennal scapes are also relatively longer. Rather than nesting in wood, it seems to prefer much more open ground. Although long known from the Channel Islands, where it is relatively common, it is a recent discovery on mainland Britain, when a nest was found in a brick wall near the Regent's Canal in central London (Smith & Williams 2008). There are now several known localities in the capital, including the recently demolished Heygate housing estate at Elephant and Castle in south London, where I found a nest under a broken concrete bollard on 13 August 2013. At the time I confused it with *L. brunneus*, which was nesting in old plane trees on the estate, but it was very clearly this new species. By the time I had realised my error the site had been razed and a return visit was impossible. In 'natural' habitats, it mostly nests under rocks and crevices in the soil, the colonies comprising a few hundred workers and a single queen. It feeds on small invertebrates and visits aphids for honeydew.

Lasius flavus (**Fabrius 1781**)

Worker 2.2–4.8mm. *Queen* 7.2–9.5mm. *Alates* August to October.

This ant is completely clear yellowish orange. It is the common meadow ant, famous for making large domed earth nests in pastures and downlands. It is one of the commonest and most widespread ants in the British Isles, occurring in a huge range of open sunny habitats, including gardens, parks, woodland clearings, commons, meadows and chalk hillsides. It occurs throughout Britain, most commonly in south and central England, and in Wales, but is also recorded from scattered localities in northern England and well into northern Scotland. In Ireland it occurs around Galway, Cork, Wexford and Waterford. Colonies comprise many thousands of workers (100,000 quoted by Macdonald 2013) and one or a few queens. In suitable meadowland, densities of 3,000–7,000/m^2 are reported. This ant is almost entirely subterranean, and no foraging takes place above ground. Its food comprises secretions from at least 30 species of root aphids, and the aphids themselves, as well as other small subterranean invertebrates scavenged or predated. Discovery of seemingly very large specimens of *L. flavus* nesting above ground in friable wood mould in several old fruit trees in Gloucestershire

Workers of *Lasius flavus* manoeuvre a cocoon in readiness for emergence. The huge size disparity suggests it will be a queen.

(Alexander 2007) raised the possibility that a new species might have been discovered – *L. citrinus* (Emery 1922). This is a temporary social parasite in *L. brunneus* nests, and is widespread in France and Spain; it has longer and narrower funicular segments than *L. flavus*. However, these turned out to be just very well nourished *L. flavus*, which contrary to received wisdom was not marshalling root aphids here, but was feeding on protein-rich click beetle grubs and other invertebrates in the rotten wood. The ant woodlouse *Platyarthrus hoffmannseggi* Brandt was also recorded above ground for the first time.

Lasius myops Forel 1894

This species is virtually identical to *L. flavus*, of which it was previously considered a variety or subspecies; it has recently been elevated to full species status by some authorities. Small workers with small eyes (and low number of ommatidia) have been referred to by this name. Seifert (2007) gives a complex formula for calculating eye proportions: (maximum eye diameter + minimum eye diameter)/(head width including eyes + head length excluding mandibles); a ratio of 0.114 $\pm$ 0.010 indicates *L. myops*, a ratio of 0.150 $\pm$ 0.009 indicates *L. flavus*. The scape is also supposed to be slightly shorter in proportion by a similar calculation: (twice scape length)/(head width including eyes + head length excluding mandibles), with a ratio of 0.780 ± 0.019 for *myops* and a ratio of 0.836 ± 0.019 for *flavus*. Obviously some fairly sophisticated microscopy, not to mention a calculator, is required for this. In northern Europe, these smaller workers are reported to occur in more southern sandy lowland heath sites (Collingwood 1979); indeed Lebas *et al.* (2019) suggest *L. myops* only reaches as far north as northern France. A more detailed understanding of its British distribution will only appear as myrmecologists take on board this nomenclatural revision and records start to accumulate. Watch this space.

Lasius fuliginosus (**Latreille 1798**)

Worker 4.0–6.0mm. *Queen* 6.0–6.5mm. *Alates* May to October.

This unmistakable ant is a shining glossy black, though the legs are slightly brownish. Pubescence is very scant, with a few scattered longer hairs over the alitrunk. The large head is broad and heart-shaped with broadly rounded hind corners. This conspicuous ant makes distinctive carton nests in old rotten trees and stumps. It's a temporary social parasite in the nests of *L. umbratus*, which itself establishes its colonies in nests of *L. niger*, *L. flavus* or *L. brunneus*. Mated queens are adopted by a host colony, but after the host queen is eradicated the colony gradually converts to wholly *L. fuliginosus* specimens. Nests comprise several hundred workers and usually one queen, but also sometimes several. The carton is made by mixing

Workers of *Lasius fuliginosus* (attending a bark-dwelling aphid), showing the large broad heart-shaped head and shiny black body.

chewed wood fragments with saliva and comprises tough papery galleries and chambers which appear dark brown to blackish. This ant occurs throughout much of England and Wales, with a few scattered records from Scotland and southern Ireland. Despite its large (for *Lasius*) size and strong coloration, this is not a commonly observed species and seems easily overlooked. The complexities of its colony foundation may give it a fragile ecology – Pontin (2005) suggests that the succession from pioneer *L. niger* to *L. fuliginosus* colony can take 20 years. There may have been some reduction of its geographic range in north and central England.

Lasius meridionalis (**Bondroit 1919**)

Worker 3.5–5.0mm. *Queen* 7.0–8.0mm. *Alates* August.

This is another clear yellow ant, and it is superficially very similar to the common *L. flavus*. The flattened scapes and tibiae distinguishing it from *L. umbratus* can sometimes be difficult to appreciate. The funicular segments are distinctly longer than wide. In the queen, the top of the petiole scale is flattened when viewed from the front. This is a rather local species of dry heaths where it nests in rotten logs and stumps. Most records are from the coast and Brecklands of East Anglia, south-east England, lowland heaths of Surrey and Hampshire, and coasts of south Wales. It is a temporary social parasite, a queen (or several) usurping a colony of *L. psammophilus*.

Lasius mixtus (**Nylander 1846**)

Worker 3.5–4.5mm. *Queen* 6.0–7.5mm. *Alates* August and September.

A yellow or slightly brownish ant, superficially similar to *L. flavus*. In the queen the top of the low petiole is broadly indented. This is a widespread species, but is not common, and although most records are from south and central England it is also found up into Scotland and in Ireland. It is generally regarded as being one of our least thermophilic species, and it makes nests deep under stones and logs in woods, pastures and rough grassy places. It is a temporary social parasite, taking over nests of *L. flavus* and probably other *Lasius* species. Most foraging is from root aphids in or near the nest.

Lasius neglectus **Van Loon *et al.* 1990**

Worker 2.5–3.5mm. *Queen* 4.0–5.5mm. *Alates* June to August.

This is a small black or dark brown ant with slightly paler legs and antennae. It is very active and superficially similar to the common black garden or pavement ant *L. niger*. It is smaller, and foraging trails up plants and tree trunks to collect aphid honeydew are much more densely populated. It is a native of Asia Minor, and was first found in Britain in the gardens of the National Trust's Hidcote Manor, Gloucestershire, in 2009 (Fox 2010). Many were found in an electrical conduit, and this species is reputed to be attracted to electricity, earning it the some-time popular name 'electric ant' in the tabloids. Queens are winged but do not appear to fly, so range expansion is slow; however, this species seems to have been transported about in horticultural material and is now known from several widely spread localities in England. Colonies have multiple queens and neighbouring nests do not recognise each other as 'foreign', so intermingle to form supercolonies. In Britain it has (so far) only been found in parks and gardens, and not in more natural habitats. In the last 20 years this ant has become widely introduced into many European countries, and it is still spreading. Large numbers of very small black ants patrolling garden plants might easily be this new species. Nests contain numerous queens and many hundreds to thousands of workers.

Lasius niger (**Linnaeus 1758**)

Worker 3.5–5.0mm. *Queen* 8.0–9.0mm. *Alates* July and August.

This common ant is all over greyish black or brownish, although the alitrunk is sometimes vaguely paler. The legs and antennae are pale to dark brown. This is perhaps the ant with which people are most easily familiar, although since the discovery of *L. platythorax* (below) easy identification is no longer clear-cut. The most straightforward character is the pubescence on the clypeus, which is dense and even in *L. niger* but sparse in *L. platythorax* (see the images of Orledge 2005). The clypeus itself is slightly more convex in *L. platythorax* than in *L. niger*. In *L. niger* the hairs on the alitrunk are more even in length, whereas *L. platythorax* has some longer hairs standing out from the rest. This species is recorded throughout almost all of Britain and much of Ireland, although it seems commonest in southern and south-east England. It makes secret nests under rocks, stones or logs, or in cracks in the soil, in gardens and parks, and is commonly the ant seen emerging from small gaps between flagstones along pavements, roads and walkways. Nests have a single long-lived queen and many hundreds of workers – up to 10,000 recorded. Its large mating flights in July or August are regularly commented on in the media. This is the species of towns and gardens, usually nesting in dry and sunny

The bog-standard black ant *Lasius niger.*

sites, while *L. platythorax* seems to occur in more natural habitats like woodlands and marshy places. It is an active and aggressive species, patrolling shrubs and herbs after aphid honeydew, and a regular invader of kitchens in pursuit of spilled sweet stuff.

Lasius platythorax **Seifert 1991**

This species is almost identical to *L. niger*, and distinguished by the subtle characters discussed above. The queens have the alitrunk more flattened dorsally than in *L. niger*, possibly an adaptation to nesting in the more confined and narrow gaps in logs and stumps. It is thought to be more cool- and shade-tolerant than *L. niger*, avoiding human habitations, and is usually not found in gardens. *L. platythorax* is definitely recorded from England, Scotland, Wales and Ireland, and is probably as widespread in the British Isles as *L. niger*, but since its separation from that species has only recently been formalised, most default *L. niger* records now need to be re-examined.

Lasius sabularum (**Bondroit 1918**)

Worker 3.5–5.0mm. *Queen* 6.0–7.5mm. *Alates* July to October.

This is a yellow to dirty orange or brownish ant, very similar to *L. mixtus* and only recently confirmed as a British species. Its status in the British Isles is as yet unclear, but it is recorded widely, if disjunctly, across southern England in parks, gardens, meadows and agricultural land.

Lasius umbratus (**Nylander 1846**)

Worker 3.8–5.5mm. *Queen* 6.8–8.0mm. *Alates* August and September.

This clear yellow ant is superficially very similar to *L. flavus*. Distinctions from *L. mixtus* are subtle and sometimes difficult to appreciate. The funicular segments, especially 2–4, are only slightly longer than wide. The body, including the antennae and legs, is covered with close silvery pubescence. It is widespread throughout England and Wales, but not common. It is more shade-tolerant than *L. mixtus*, often nesting in rotting wood in damp woodland. It is a temporary social parasite, taking over nests of *L. flavus* or *L. niger*, and possibly other species. Most foraging is on root-feeding aphids in and near the nest.

Camponotus ligniperda (**Latreille 1802**)

Worker 6.0–14.0mm. *Queen* 16.0–18.0mm. *Alates* May to July.

In this ant the head is black, the alitrunk, petiole and fore face of the gaster are bright red, and the rest of the gaster is black, moderately shining. The legs are reddish brown. Petiole and gaster have sparse but long outstanding hairs. It is one of several species that are widespread in southern Europe and Scandinavia, occurring throughout northern France and Belgium. It is sometimes brought into the British Isles on imported timber, but no colony has yet become established. I've often wondered what were the strongly coloured and very active ants I found under scraps of loose bark still adhering to planks of cut timber being unloaded from the cargo ship *Con Zelo* at Newhaven Docks in

Camponotus ligniperda and young brood.

the summer of 1976. I worked there as a teenager between finishing secondary school and going to university. This rather rusty white ship plied a regular trade between England and France, appearing at the docks every few days, usually bringing in whole tree trunks, cut into planks, but then steel-banded back up again and often containing small areas of bark, rot cavities and various bits of detritus. I now suspect that the ants were this species. The casual stevedoring on the North Docks was tedious, dirty and tiring, and although I noticed the ants during the boring waits between fork-lift truck-loads of sawn timber coming from the quayside cranes, I had no collecting paraphernalia about my person and failed to gather a specimen. I've never made that mistake again. Other *Camponotus* species are known to turn up occasionally, imported live with horticultural material; an unidentified species was found infesting a palm tree brought in from Argentina for the Hampton Court Flower Show on 8 July 1999 (Andrew Salisbury, personal communication). The large (7.0–15.0mm), dark, heavily built *C. herculeanus* (Linnaeus 1758) occurs widely in Scandinavia and central Europe and might be a potential future colonist, especially since this is thought to be one of the most widely spread ants on Earth, with an apparently natural Holarctic distribution (Schär *et al.* 2018).

Polyrhachis lacteipennis Smith 1858 (formerly *P. simplex* Meyr 1862)

Worker 4.5–7.0mm.

This ant is black, the body finely punctured and granulate, giving a matt appearance. It is also very spiny – one pair of short spines emanating from the pronotum, one longer pair on the propodeum which are reflexed and angled out and forwards, and a very long upwardly curved pair from the corners of the petiole. These large ants make carton nests at the base of trees, using saliva to cement twigs together. They occur in India, Nepal and the Middle East. This is not a British species, but numerous specimens, complete with nest, were accidentally brought to the Chelsea Flower Show in a large date palm transported from the United Arab Emirates. Several specimens were collected from the tree on 23 June 2001 (Halstead 2001), and still reside in the reference collection of the Royal Horticultural Society (Andrew Salisbury, personal communication). This shows that almost any ant from anywhere in the world could turn up almost anywhere.

Evolution of ants

Although ants form a discrete and self-contained group among the huge insect order Hymenoptera, they are obviously closely related to bees and wasps. There are many seemingly nearly intermediate forms, or groups which appear to share some common characteristics. Winged, large-eyed male ants look rather waspish, and several ostensibly wasp groups have wingless, and therefore rather antish, females. These similarities become even more evident when examining the geological record. Nevertheless, clearly recognisable ants have been around for at least 100 million years. We have exquisitely preserved fossils in amber to prove it.

Amber is fossilised tree resin. As a small boy climbing the large old cherry trees in my grandad's Kent orchard I was fascinated by the oozing excrescences of clear, yellow, sticky, resinous sap dribbling from cracks and folds in the bark. With the consistency of old chewing gum, but the appearance of molten glass, it seemed like treasure and I broke bits off to take back to show and tell at the dinner table. I was fascinated by the tacky sensation on my hands, which no amount of soap seemed able to wash away. I suspect my mum was forced to attack my disreputably dirty fingers with a scrubbing brush and Swarfega, and probably cursed me when she found bits of resin immutably welded into my trouser pockets on wash day.

I wasn't the only one to find the resin attractive; occasionally a beetle, fly or small hymenopteran would get stuck to it, like fly-paper, and any amount of struggling just seemed to get it more firmly entrapped in the enveloping goo, to the insect's ultimate demise. I didn't realise it at the time, but this was a grisly scenario that had been playing out for aeons. Although the oldest ambers are 320 million years old, it is from around 150 million years ago (MYA), with the appearance of flowering plants, that small creatures from

OPPOSITE PAGE:
Tiphia femorata, an ant-like wasp. Ants and wasps share a common ancestor that lived about 200 million years ago.

mites to mice have been getting stuck in similar tree exudations. It is impossible to be sure which antediluvian trees were doing the exuding, but theoretical possibilities include ancestors of: (a) modern New Zealand kauri trees (*Agathis* species); (b) South American/ African *Hymenaea* trees (legume family); and (c) umbrella pines (family Sciadopityaceae), now only one species found in Japan.

The trees that dripped their sticky sap in the Cretaceous and Eocene are now extinct, as too are the insects and other small critters that became entombed in the resin. But for some reason that resin has survived to bring its beautifully preserved contents down to the present day. Tree resins are often produced as a response to physical damage, and they are thought to block up the jaws of wood-boring insects trying to get under the bark, and to scab-over open wounds through which diseases might enter. They contain complex mixtures of highly aromatic and volatile substances such as alcohol, ether and light hydrocarbons, within a base of larger-molecule cyclic terpenes. As the smaller molecules evaporate away, the resin nodule thickens and solidifies; the terpenes polymerise, chemically joining to form an almost crystalline mass, and creating a tough natural substance very like plastic – a solid plug barrier against insect attack or fungal infection. Insects and other small animals on the tree trunk easily get snagged in the sticky resin, and copious waves of the stuff engulf them until they are suspended in the glistening globules. Today, similar resins, called copal, from modern trees like the South American *Protium copal* and East African *Hymenaea verrucosa*, continue to trap insects, and lumps of copal as large as beachballs are well known.

The chemical polymerisation continues, even if the resin nodule drops to the ground, or the tree dies and falls. The aromatics go on to dissipate and the terpenes continue to knit together until they become effectively inert. Now they become buried, first under leaf litter, then under soil, silt and rock laid down over them for millions of years. Eventually, as tectonic movements shift and twist the very crust of the Earth, they become uncovered by wave action on the ocean floor and are washed up on the beaches (amber's specific gravity is 1.04–1.10, just slightly denser than water), where curious humans pick them out among the sand and pebbles and wonder at the sun glinting through the lustrous transparency, and their preserved treasures within. The chemicals in the sap obviously had similar defensive anti-insect, antifungal and antiseptic activity to modern resins, and they prevented the insect inclusions from rotting. Instead, the animals'

Unnamed ant in amber. Although unidentified yet, this is clearly a formicine worker, with the single-segmented petiole and 12-segmented antennae.

body structures have been preserved in their original often pristine state. Amber inclusions include virtually all known insect groups, especially those that would normally be expected to crawl up and down tree trunks. Ants, not surprisingly, feature heavily. One of the oldest ambers appears to be from the Dolomite Alps of northern Italy (about 230 MYA), although Burmese (about 100 MYA), Baltic (50–30 MYA) and Dominican (from about 20 MYA) are the best known and most studied ambers with fossil insects inside.

Amber has long been valued and combined into lustrous jewellery. Its precious (to humans) nature is best exemplified by the famous Amber Room erected in the Catherine Palace in Tsarskoye Selo near St Petersburg during the early 18th century. Backed by gold leaf and mirrors, the 55 square metres of amber panels were a wonder of the world until they were looted by the Nazis in 1941. Today amber jewellery is still highly prized, I've bought it myself. But now, inclusions like insects, spiders, lizards, frogs, flowers, leaves, or more recently dinosaur feathers and bones, have added a premium to an already luxury item. A quick internet search shows that ants preserved in amber are widely available from dealers. I was tempted by several in the £15–30 price bracket. The ants were all tiny, 2–4mm, but with low monetary value they were unlikely to be fakes. However, I was not tempted by the mating 'queen and drone' for sale from a US website at $49,900. If memory serves, this is the same item that I came across more than 10 years ago when

Left: winged queen formicine ant in Eocene amber.

Right: myrmicine ant in Dominican amber. The double-segmented petiole/postpetiole is easily visible.

I was researching 'the most expensive insect' (Jones 2010), and which was then on offer for $100,000. No matter how rare it might be for such an event to get fossilised, this seems to be a fairly arbitrary sum, even if now knocked down by 50%.

Fossil ants are mostly from Baltic and Dominican amber from 40–37 and 19–16 MYA respectively, by which time modern lineages had become well established. On the whole the fossil record for ants is quite good; an oft-quoted statistic is that there are about as many described fossil ant species as there are dinosaur species – approximately 730 from 67 deposits worldwide (Barden 2017). Given the huge literature and even huger interest in dinosaurs, it's tempting to imagine that ant prehistory is well understood. However, there are some important caveats. As for any group of organisms, great care has to be taken when interpreting the fossil records. Firstly, fossilisation is an extremely uncommon and unlikely event. Virtually every organism that dies decays away – scavenged, broken, weathered, rotted down to nothing. Flimsy invertebrates are just about the least likely things to get petrified. Fossils are also highly unrepresentative of any fauna. Ants trapped in amber tend to be small (up to 10mm long) species that probably foraged on the trees that leaked the stuff. Small, narrow, nearly eyeless subterranean species are rare in amber, and are usually the winged sexual forms that have presumably become trapped during mating flights. Imprint fossils in rocks are formed in silt layers, mostly in lakes, and are mainly preserved winged males and queens that fell into the water, also during mating flights. These fossils are only discovered by splitting the rock layers, an intrusive process that always has some damage associated, and by which small individuals are easily overlooked.

Miracles in stone – why fossils are so rare

Dinosaurs have captured the popular imagination, but this has led to some completely silly and imaginary ideas about fossils. Fossils are real, but they are so rare that they do, actually, verge on the mythical. I casually wondered out loud (well, on social media) what might be the chances of something getting fossilised. This seemed like quite a straightforward question to me, but I soon found out that nobody was willing to offer up any firm calculations. Taphonomists (fossil palaeontologists) politely responded to my question, but I could almost detect a mocking tone, as they claimed that 'putting odds on it is impossible'. Having thought about it a bit more, I'm tempted to agree with them. Nevertheless, having set the metaphorical ball of fiery curiosity rolling through my brain, I here offer some back-of-the-envelope calculations.

Firstly – how many ants have ever been available for fossilisation? Starting with that statistic already used on page 22, Williams (1960) estimated 10^{18} insects on the planet at any given time, of which say 1% were ants; this would give us 10^{16}. Assuming some sort of conformity with modern faunas, and allowing each ant to live one year, the last 100 million (10^8) years might therefore have produced $10^{16} \times 10^8 = 10^{24}$ ant individuals having ever crawled on planet Earth.

Secondly – how many fossils are known? It turns out that about 750 fossil ant species are described from, say, 10,000 ant fossils. At the final count this does not seem like very much at all.

Finally – what proportion of fossils have we found? Of course humans have only discovered a tiny fraction of the fossils buried in the soil. This is where the calculations and guesswork start to get a bit fanciful. For the sake of argument, I suggest that humans have only discovered 1 in 100,000 – that's 0.001% of ant fossils. This is a figure I admit to having arbitrarily plucked out of the air. I dared somebody to disagree with me and they readily did, coming back with suggestions that my estimate could easily be a million- or billion-fold wrong.

Nevertheless, ploughing on, if 10,000 fossils represents 0.001% of the total, that total might be 10^9 fossil ants scattered around the globe. This would still give the chance of fossilisation as something like $10^{24}:10^9$, or in more conventional usage $10^{15}:1$. These are incomprehensibly small odds.

Given that the chance of winning the UK's weekly National Lottery is reportedly $4.5 \times 10^7:1$ (45 million to one), the chances of an ant becoming fossilised are about on a par with the same person winning the lottery two weeks running. Given that 32 million people are reputed to do the lottery each week, my increasingly tenuous grip on mathematics suggests that it might take 613,278 years for them all to have a win. For someone to win twice in a row seems mathematically (and morally) untenable. For an ant to become fossilised must be equally dubious.

It's that unknown how-many-more-fossils-are-left-in-the-ground part of the equation that gives me most anxiety, and about which most people have given me stick. But even if my guesstimate is out by a billion-fold the odds are still vanishingly small, even if we can't put a realistic number on them. Really, in any sensible world, fossils shouldn't exist – each and every one of them is a tiny miracle of geological chance. Perhaps I should have shelled out $49,900 for that mating ant pair in amber.

Bees, wasps … and the rest of them

Identifying 'recent' ants from the last 100 million years or so is at least eased by being able to fit them into modern classification schemes. It's the earlier evolutionary history of where proto-ants came from that needs to be clarified in the deep fogs of geological time. And this is where quite a lot of guesswork (interpolation, in scientific speak) is required.

According to the masterful work of Grimaldi & Engel (2005), the Hymenoptera are the most difficult group of the Holometabola to place into some sort of overall insect phylogeny. Although the Hymenoptera form an obvious and distinct group, their relationships with other insect orders are problematic. They share a number of primitive traits with each other, but show little overall resemblance to other insect lineages, making it very difficult to guess who their nearest relatives are. At the moment the Hymenoptera are considered to be allied to the Panorpida. This large group includes the scorpion flies (Mecoptera), butterflies and moths (Lepidoptera), fleas (Siphonaptera), flies (Diptera) and caddis flies (Trichoptera). The similarities, though, are extremely slight, and based on some very obscure characters such as reduction to a single claw on the larval leg, labial silk glands in the larva, and a sclerotised sitophore plate in the cibarium of the adult mouthparts – a tiny structure involved with creating suction during liquid feeding. The prothorax is also reduced. Comparative DNA studies between extant species seem to support this association too. Examining fleas and flies or moths and caddis, it is easy to appreciate that they share a close evolutionary descent. Their structural similarities are easy to observe under the lens and their relatedness is intuitively obvious. But the origins of Hymenoptera are a source of constant bafflement.

Insects are thought to have evolved from some six-legged arthropod ancestor approximately 400 MYA around the Silurian/Devonian boundary. Larvae, suggesting metamorphosis and the appearance of the Holometabola (insects going through the 'complete' metamorphosis of pupa/chrysalis stage) are known from 300 MYA. The Hymenoptera are likely to have split very early on from the remaining holometabolous orders, and in many schemes they are given their own suborder status – the Hymenopterida – to get over the fact that they seem to have gone it alone for so long.

The earliest obvious Hymenoptera fossils are Triassic (250–200 MYA) compression fossils of sawfly wings, easily recognisable because of their distinctive venation. Today the sawflies are usually regarded as basal, primitive groups within the Hymenoptera. With plant-feeding caterpillar-like larvae, a rather stubby abdomen broadly connected to the thorax and lacking any sting, they do not look at all like the waspish majority of the rest of the order. Nevertheless they have the key defining features of the Hymenoptera – hamuli hooks for attaching large front and small back wings together, a velum membrane and spur on the front tibia used for antennal cleaning, and the strange chromosome-number sex-determination system that is haplodiploidy (see page 194).

The evolution of the wasp waist was a momentous event, conferring a manoeuvrable flexibility on the vast majority of the Hymenoptera and suggesting the evolution of a sting and the predaceous or parasitic lifestyle in the group now called the Apocrita about 190 MYA. As mentioned in Chapter 2, fusing the first abdominal segment onto the thorax to create the alitrunk/mesosoma allowed the petiole to develop, and now the weaponised sting can be angled in almost any direction to subdue prey and/or lay eggs. Today the parasitoid groups are the most speciose and diverse of all the Hymenoptera, ranging from the large Ichneumonoidea to the tiny Chalcidoidea and Proctotrupoidea. They inject tiny eggs and paralysing or killing venom through a sometimes unfeasibly long and fragile-looking ovipositor. Among them emerged a line which retained the sting, but which did away with any long egg-laying tube. Although venom still passed down a narrow hypodermic tube, it was considerably shortened compared to parasitoid lineages, and usually fully retained in short telescopic form inside the tip of the abdomen. Meanwhile, the egg passes through a hole near the base of the sting. This is the most familiar and famous group of the Hymenoptera – the bees, wasps and ants – collectively called the Aculeata after the sting (the aculeus) which has made humans so wary of them.

The earliest evidence for the aculeates is a series of compression fossils from central Asia, 150 MYA. These small ant-like wasps in the now extinct family Bethylonymidae are similar to the living group Bethylidae which we met in Chapter 2 (sorry, still no common English name). They appear to have had a short sting and 12 or 13 antennal segments – similar enough to modern aculeates, but not quite evolving the 12 female/13 male antennomere formula we know and love today. From here on down, almost all aculeates are really just 'wasps'. A few slightly hairier groups developed a taste for pollen and nectar, and these are now collectively, but rather arbitrarily, called bees (families Apidae, Andrenidae, Halictidae etc.).

Antness appears in the fossil record about 140 MYA with the emergence of the false ants – Falsiformicidae and Armaniidae. These two extinct groups of not-quite ants show ant-like wing venation, a broad petiole giving a slight waist, geniculate antennae (although with scape shorter than in most modern species), and roughly queen-like females, but they lack the metapleural gland. Myrmecologists now take this gland, opening at the lower rear corner of the propodeum near to where it articulates with the petiole, as one of the key features of the Formicidae, and it is found in no other insects. In modern ants it exudes a range of antimicrobial and communication secretions. It is, however, lost in many arboreal or social parasite species and is usually absent in male ants. Falsiformicids and armaniids are known exclusively from imprint fossils; these are often poorly preserved and sometimes

difficult to comprehend. Visualising the tiny metapleural gland orifice is often a task beyond practicality or possibility. Both families have been rearranged several times under different interpretations of ant prehistory, but the general consensus now is that they are sister groups to ants, rather than actual ants. Their fossils show wasp-like mandibles which are either bidentate or have just a single apical tooth, and they share many features with the conglomerate superfamily Bethyloidea which includes the Bethylidae parasitoids mentioned earlier, rubytail 'cuckoo wasps' Chrysididae, which lay their eggs in the nests of other wasps, and Dryinidae, which have wingless ant-like females that parasitise leafhopper nymphs. Unfortunately, the decisive feature that unequivocally identifies ants – worker eusociality – is not preserved in any of these fossils.

True ants arrive on the scene

Although no true ants are recorded before the Albian stage of the mid-Cretaceous (113–100.5 MYA), the transition from ant-like wasps to wasp-like ants is likely to have occurred something in the order of 100–150 MYA. Seminal among discoveries was that by Wilson *et al.* (1967), who described *Sphecomyrma freyi* from New Jersey amber reckoned to be 92 million years old. Although it had double spurs on middle and hind tibiae, like modern wasps, and only two teeth on its short stout mandibles, its narrow petiole, long flexible funiculus, and metapleural gland, much more easily visible through the polished amber, firmly landed it true ant status. Later, Perrichot *et al.* (2008) reported two wingless worker females preserved together in the same piece of amber and concluded that they were, indeed, eusocial. Several other amber fossil species have since been described in the same extinct subfamily Sphecomyrminae, including the first species with both queen and worker castes, *Haidomyrmodes mammuthus*. These were all likely foragers on the coniferous trees that produce the amber resin from wounds in trunks and branches. Again, the antennal scapes were shorter than in most modern ant species, but there is no mistaking that these are real ants, and they are otherwise very similar in general appearance to modern species. Their relatively frequent occurrence in both French and Burmese ambers suggests that this group was widespread across Laurasia – the northern hemisphere North America/Europe/Asia supercontinent following the break-

ABOVE: The oldest accepted true ant fossil, *Sphecomyrma freyi* (left), and the earliest known formicine, *Kyromyrma neffi* (right), both discovered in New Jersey amber from about 92 million years ago.

up of Pangea. Conversely, the oldest ants known from Gondwana (the southern hemisphere South America/Africa/India/Australia/Antarctica supercontinent) are later – Botswana, about 90 MYA. They are diverse, thought to be Formicidae, but of 'uncertain' subfamilies. From this observation, combined with the apparent absence of the nearly-ants Armaniidae from several rich fossil deposits in the area, it is tempting to suggest that the earliest diversification of ants occurred in Laurasia, in the north, but … absence of evidence is not quite evidence of absence. Although no sphecomyrmines exist today, the wide array of their fossils shows that they were a morphologically diverse group, implying that their behaviours and life histories were also varied.

The earliest known formicine ant, *Kyromyrma neffi*, was also described from 92-million-year-old New Jersey amber, and again the amber allows perfect visualisation of the acidopore. Other modern subfamilies all seem to have existed from about this time too. That same 92-million-year-old New Jersey amber has yielded several other specimens including the earliest ponerine, *Brownimecia clavata* (now in its own subfamily Brownimeciinae). The earliest dolichoderine *Eotapinoma macalpini* is known from Canadian amber (79–56 MYA). Despite this, early ant fossils remain scarce, comprising only 1–2% of recorded fossil insects, suggesting that, although present, ants had not become the dominant terrestrial arthropod force they are today. Indeed, the absence of ants from Lebanese (125–110 MYA), Spanish (113–100 MYA) and Brazilian amber (120 MYA) has been used to argue that they were not present, or at least not significant, on Earth at that time. They gradually started to increase – figures given by LaPolla *et al.* (2013) suggest that ants make up about 13% of known insect fossils in the Eocene (50 MYA), 20% in the Oligocene (35 MYA) and 30% in the Miocene (18 MYA). For anyone wanting to go deeper into the fossil history of ants, extensive and informative tables detailing ant-bearing fossil deposits are given by LaPolla *et al.* (2013) and Barden (2017).

During the Eocene, 55–35 MYA, the confusingly, and I think unfortunately, named Formiciinae (not to be confused with the current Formicinae) appeared, then vanished. Some of these were the largest ants to have ever lived, with queens of *Titanomyrma lubei* easily reaching 60mm long, nearly twice the length of any standard living ant today (although some queens of *Dorylus* driver ants are reputed to reach 50mm), and with a wingspan of 130mm to rival modern dragonflies. In flight they must have looked like modern-day hornets. Large ants are perhaps more likely to be fossilised, or their fossils more likely to be discovered. An Eocene find in Denmark from about 55 MYA revealed 101 isolated body parts of males and queens of a species of giant ant called *Ypresiomyrma rebekkae* – perhaps the earliest evidence of a mating swarm (Rust & Andersen 1999). Baltic Eocene amber (42–37 MYA) has produced a long series of important ant fossils, including the first evidence of major and minor workers in *Gesomyrmex boernesi* and a *Pheidologeton*

Titanomyrma, from about 50 million years ago, were some of the largest ants to have ever lived.

species, and a putative ergatoid queen of *Plagiolepis klinsmanni*. Several fossil ant species are associated with aphid inclusions in the same amber pieces, suggesting the familiar trophobiotic herding behaviour of so many modern ants had evolved at this point. Indeed, one winged queen of *Acropyga glaesaria* is still holding a mealybug in her mandibles. Today, extant *Acropyga* feed solely by tending the obligate mealybugs inside their nests, and during nuptial flights each queen makes off with a 'seed' mealybug in her jaws, with which to found a new colony in her new nest.

The ants diversify and start to attract attention

A nice parallel of increasing ant abundance and diversity also exists in the fossils of ant-mimicking insects, with several ant-like longhorns (Cerambycidae) and rove beetles (Staphylinidae) known from Dominican amber (19–16 MYA), suggesting that ants had become significantly potent models for mimicry by then (Poinar 2010). A parasitic ant mite, *Myrmozercon*, was found attached to the head of a specimen of *Ctenobethylus goepperti* in Baltic amber (42–37 MYA). Even earlier, Cambay amber (52–50 MYA) had revealed the first known ant-nest beetle of the rove beetle subfamily Clavigerinae (Parker & Grimaldi 2014).

For queens, workers, majors, minors and ant-nest beetles to occur, it seems obvious that there must have been ant colonies, and they must have been living in nests, but fossil evidence is scant, and difficult to interpret. The oldest is reckoned to be a series of nine nests with conical entrances, found together in Upper Cretaceous (80–70 MYA) mudstone and sandstone deposits in Utah (Roberts & Tapanila 2006). Cross-sections of the globular relics reveal tunnels and chambers consistent with modern subterranean ant nests where species nest in moist soil, including some shoring up and backfilling after an initial silt-laden flood, before the final overwhelming 'overbank flooding event'. Since no fossil insects were found associated with the mounds, interpretation has to be circumspect, and such trace fossils (ichnofossils) are named in the absence of known makers. These nine nests are order-neutrally named *Socialites tumulus* (= 'mound-building associates'), and although termite construction cannot be completely ruled out, the fact that the tunnels appear not to be reinforced by a lining of saliva- or faeces-based cement is taken as good evidence that the builders were ants. Elsewhere, ant pupae are well known in amber, and can be identified because the resin renders the silken cocoon transparent so the chrysalis inside can be visualised. How the pupae became entrapped in tree-exudate amber is an unanswerable question. Being immobile, and usually hidden inside a nest, one can only assume that they were being transported by workers to escape from a disturbed nest (flood, predator excavation, invasion by other ants), or that they were some itinerant species, like army ants, moving from one bivouac to another. The earliest known ant-colony fossil, of 438 individuals, complete with major and minor worker head sizes, larvae and pupae, was described by Wilson & Taylor (1964) as a new species *Oecophylla leakyi* from the Miocene, about 20 MYA.

Ants were also plentiful enough to be presumably under attack from specialist predators, since an ant bug *Praecoris dominicana* (Hemiptera, Reduviidae) is known from this time (Poinar 1993). Modern relatives of this species have a gland on the underside of the abdomen that attracts and paralyses ants. Shortly after imbibing the gland secretion an ant becomes lethargic and is easy prey to the sharp stout proboscis with which the bug sucks out the ant's innards. Long stiff hairs on antennae and legs offer protection against ant jaws. The closely related *Proptilocerus dolosus* was found in Baltic amber with two shrivelled bodies of what were very probably its ant prey – *Dolichoderus tertiarius*. The extant bug *Ptilocerus ochraceus* continues to feed on modern ants in the same genus.

Likewise, a butterfly caterpillar from Dominican amber is very similar to modern metalmarks (family Riodinidae); many tropical riodinids live with ants in a similar way to the blues discussed in Chapter 8. A pair of large glands on the penultimate body segment provide nutritious secretions that the ants lick, a fringe of long bristles around the caterpillar may offer protection against nipping ant jaws, and paired bumps behind the head are either vibratory or produce chemicals that influence the ants' behaviour.

By the Eocene (about 56–34 MYA) ants had become a significant enough part of the world's ecology that they had specialist mimics, predators, parasites, nest-invaders and inquilines ('guests'). There is some debate about whether there was an explosion of diversity mirroring the rise of angiosperm flowering plants at about this time. The major groups of plant-feeding beetles (leaf beetles, longhorns and weevils) show a good correlation and became highly speciose as angiosperms diversified, but in these almost entirely herbivorous groups there is a high host-plant specificity that does not hold true for the far more generalist ants. Certainly increased flowering plants would have provided more nectar, seeds, honeydew from increasing aphid and plant bug species, and potentially more prey associated with the new plant forms, but even today ant diversity is not intimately linked with plant diversity. Diversifying plant architecture and correspondingly more complex leaf-litter layers may have had some small effect though.

The recent technique of DNA sequencing has also been brought to bear on trying to work out a tree of evolutionary descent for all modern ant subfamilies. A review by Moreau (2009) gives a phylogeny for 21 extant subfamilies, and these more or less follow earlier ideas of ant subfamily relatedness based simply on morphological characters, although with some reshuffling. They also allow calculations to be made regarding molecular evolution on the assumption that the random base-changes in the DNA sequence occur at a more or less regular mutation pace throughout geological time. It was this technique that gave rise to the suggestion that although no fossils are known from that time, ants probably first arose at some late point in the Jurassic period, 201–145 MYA (Crozier *et al.* 1997). There have always been concerns about the potential errors when calculating by extrapolation from chemical DNA base-changes into millions of actual years, and many authors suggest that these dates are the maximum possible, and quite likely to be overestimates. Because of the fragmentary nature of the fossil record, and the intrinsic unlikeliness of fossilisation events, it is perhaps not surprising that the earliest possible fossils have not yet been found.

What does seem likely, though, as revealed in comparative DNA analysis, is that the period of greatest ant diversification happened around 110–92 MYA, when all modern ant pedigrees seem to have been established. The major radiation was in species in the now dominant subfamilies Formicinae, Myrmicinae and Dolichoderinae, all of which contain genera (*Formica*, *Atta* and *Azteca* respectively) which make highly populous nests with widely foraging workers that are more likely to get trapped in the sticky tree resins from which amber is formed.

Meanwhile, Brady (2003) demonstrated from the ants' DNA similarities that, contrary to earlier ideas of convergent evolution, army ants evolved about 100 MYA in the southern supercontinent Gondwana, and that they subsequently diverged to form the South American *Eciton* 'army ant' lineages and African *Dorylus* 'driver ant' species.

Today we know of about 12,500 extant ant species, with many more waiting to be discovered or unravelled. Trying to work out how they evolved, and around which fossil ancestors they passed by or through, is a constantly moving multidimensional jigsaw puzzle. Knowing the definitive answer is not necessarily the point – as with all science, it's about asking the right questions. Today, an understanding of how ants are related to each other in time and space gives us a basis to understand the ant biodiversity we have now, how it fits in to the life of existing habitats and ecosystems, and how we can value and therefore preserve it for the future.

Being an everyday ant

Ant behaviour is manifold, and complex. No ant is a loner, idly wandering the world pleasing only itself; instead, every individual is part of a complex colony system and its behaviour is to some extent controlled by the needs of the colony. This is best (and everywhere) exemplified by the ant behaviour that is most often observed – foraging for food. It is, though, easily misunderstood. The cartoon stereotype of ants raiding human food, whether at the picnic or in the pantry, is accurate to only a very small degree. Ants have existed for far longer than humans have had either picnics or pantries. This modern behaviour is simply a manifestation of a long-evolved instinctive activity.

The 'well-known fact' that ants pinch human food stems from the traditional ancient knowledge that some ants collect seeds and take them back to their nests. But scavenging leftovers is just a minor part of ant nutrition for a tiny minority of species. Ants are, however, highly adaptable, and they will take advantage of whatever suitable foodstuffs they come across. And because we humans live in our own houses, we tend to notice if ants have come across something interesting in the larder. Ants are active, curious, boldly explorative, and often single-minded in their adventures across the human threshold, and because they usually work in large gangs, it's not long before their initially secretive activities are discovered.

In urban areas it is usually the black pavement ant *Lasius niger* that makes these intrusions. This is an extremely common ant species, which occurs in gardens throughout the country, and since it tends to nest in cracks in pavements, patios, garden paths and frequently in broken masonry, it is never far from the back door. The last time I was invaded, they had discovered the sticky residue around the

OPPOSITE PAGE:
Myrmica rubra workers herding a large crowd of aphids.

Although almost certainly staged for the photo agency, this is most people's standard nightmare image of ants invading their food.

bottom of the golden syrup tin on the pantry shelf, where a dribble had not been completely wiped clean. Elsewhere *L. niger* will attend discarded chewing gum, dropped sweets and ice-cream drips on the pavement. They are taking advantage of the sugary liquor resembling their natural food, aphid honeydew – more on this shortly.

There are reports of the reclusive brown tree ant *Lasius brunneus* visiting larders in old houses. This is normally an inhabitant of mouldering tree trunks in ancient woodlands, but nests are sometimes found in the large beams of old timber-framed houses and farm buildings. Other interlopers are also available. In Ashdown Forest, many years ago (early 1970s), my father and I chatted to a wood-cutting forester after he was curious about our insect nets. He spoke of how the local wood ant, *Formica rufa*, had invaded his house and how he had prevented its incursions by drawing a line with a piece of chalk across the threshold of the door. His anti-ant defence always seemed much more like witchcraft voodoo magic to me, a sort of pseudoscience that equated combatting the ants' formic acid with the calcium carbonate base in the classroom chalk. However improbable his tale seemed, it is just possible that a local wood ant nest had taken a temporary foraging route under his door – in nature nothing ever turns out to be impossible.

Domestic attacks from ants are often from non-native species; these 'tramps' have been transported around the world by human commerce, travel and imperial aspirations. In the temperate British Isles, away from the warmer tropics, they can often only survive inside heated buildings, so the nest is usually inside the fabric of the building itself, rather than the ants breaking and entering from outside. The notorious pharaoh ant *Monomorium pharaonis* and Argentine ant *Linepithema humile* are the usual culprits. The large nests, containing potentially many hundreds of thousands of workers, need a constant supply of food – which inside hospitals, schools, offices or homes can only be obtained from human stores, or rubbish bins. They will ant-handle small morsels of food or suck up any liquid sweetness they can find to take back to the inner recesses of the colony.

Elsewhere in the world, many more species of ants come indoors. This is partly because there are more species out there anyway, but also because the feeding and foraging strategies of cool-temperate British ants are subtly different to those of subtropical species. Many exotic ants are large and long-legged, and their far-ranging food-hunting explorations are perhaps more likely to tempt them under the kitchen door. Scavenger/predator behaviours (there's usually only a fine line between them) are also more prevalent in subtropical species. Seed-gathering occurs only in one uncommon British species (*Tetramorium caespitum*), and only as part of a general foraging strategy, but this behaviour is widespread in warmer climates. This is the behaviour exhibited by *Messor* harvester ants around the Mediterranean and Middle East – Messor was the Roman god of the harvest, and the name means 'he who reaps'. They are readily and easily observable as the large (and large-headed) black ants that drag seeds, grains, ovules and grass florets quite some distance across the ground back to the nest. It is, perhaps, only a short behavioural jump from collecting seeds to gathering food crumbs found scattered on the kitchen floor, on larder shelves, or under the cooker.

Explorative foraging into the house is also less likely in north European ants because by far the majority of temperate ant species form rigidly close-knit colonies with ferociously constrained territories, constantly surrounded by and abutting other colonies, thus preventing the long-distance foraging trails that they would need to find their way into our houses. In addition, the often hermetically sealed modern buildings of the developed world (with their manicured and gentrified surroundings) tend to be clinically separated from the natural habitat that wild ants inhabit, whereas in much of the developing world there are fewer and smaller physical barriers between the wild ants and the human resources they sometimes raid.

Feeding – finding, sharing and eating food

Apart from the few tramp species that have spread across the world, and which survive mostly in human habitations, most ants have their own natural nourishment. And although adult ants do need a little bit of food to give them energy, they follow the standard insect policy that most of the feeding (and all of the growing) is done by the immature stages – the larvae. So naturally, most of the food they collect is brought back to the nest to feed the brood. The Ponerinae, generally accepted as one of the most primitive ant subfamilies, are mostly predators. The uncommon and secretive *Ponera coarctata* is mainly carnivorous; it hunts and collects small invertebrates which it brings back and leaves in the brood chambers for its grubs to help themselves.

Elsewhere in the world, South American army ants, *Eciton burchelli* and others, and African driver ants, *Dorylus* species, are well known for their aggressive hunting techniques, spreading out across the forest floor and overpowering any small organism that gets in their way. They attack beetles, spiders, scorpions and caterpillars, but will also swoop on snakes, lizards, nestling birds or other larger animals which, because of infirmity or injury, cannot escape. The raids take

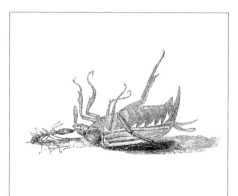

ABOVE: A very ambitious ant. A rather unrealistic vignette from *A Familiar Introduction to the History of Insects* (Newman 1841). Ants are remarkably strong for their size, but it would require a dozen to move this dead cockchafer.

RIGHT: Entitled 'insects at work', this decorative illustration from Michelet (1875) shows how ants will, indeed, visit carrion to scavenge morsels, but tales of them quickly skeletonising large corpses are exaggerations.

the form of an amoeboid pseudopodium – a semi-amorphous blob-like tendril that extends, surrounds and overwhelms. However, tales of these ants attacking dogs, farm animals or humans are probably to be taken with a pinch of salt. Despite the purported execution of criminals mentioned by 19th-century traveller Paul du Chaillu in Africa, the pegging out of human victims to be dismembered by ants is a Hollywood trope beyond truth, and the image of hordes of giant ants dragging still-screaming Soviet soldiers down into a nest burrow in *Indiana Jones and the Kingdom of the Crystal Skull* is a comic-book comedy of risible proportions. *Dorylus* driver ants have sharp cutting mandibles (*Eciton* army ants are chewers), but tales of the ants stripping cows or people to the bones within minutes are entirely fictional. Detailed studies of skeletonisation by ants are sparse, but fire ants *Solenopsis invicta* frequently visit carrion and are said to be able to strip a dead frog to the bones in about 12 hours.

Feeding prey to larvae is not always a one-way process. In the North American *Pheidole spadonia*, dismembered insect portions are placed onto hairy depressions near the mouth of the larva, which secretes digestive juices. These dissolve the fragments into a soup-like slurry which is eagerly distributed to other members of the colony.

A raiding party of *Eciton burchellii* army ants streams back to its bivouac camp in the Ecuadorian Maquipucuna Reserve. A large soldier stands guard at the edge of the trail.

Red ants, *Myrmica* species, work together in their dozens to scavenge a large hoverfly, *Scaeva pyrastri*.

In experimental measurements it took about five workers and 22 fourth-instar larvae 12.8 hours to dissolve one fruit fly carcass (Cassill *et al.* 2005). Regurgitation of liquid food by larvae is found in almost all ant subfamilies. This liquor contains an amino acid mix two or three times more concentrated than in the grub's haemolymph, the insect equivalent of blood. Thus there is a good scientific argument for regarding ant larvae as a specialised digestive caste within the nest.

It's important to mention larval haemolymph feeding here, if only to head off the rather lurid idea of so-called Dracula ants. Several (non-British) species in the subfamily Amblyoponinae (and a few outside it) indulge in drinking the body fluids of their own larvae. For many species, this is the only way that the queen gets any nutrition. She gently nips the nearly fully grown larva, puncturing its skin on its upper surface, then imbibing the haemolymph that leaks out. The larvae seemingly survive this abuse without any long-term harm. The wounds of the 'donors' heal, and some idea of the frequency of the behaviour can be gauged by monitoring the scarring level of the brood. Rather than 'blood-sucking', with its gothic horror connotations, this behaviour might best be described as non-destructive or non-fatal sub-cannibalism.

Prey (or carrion) is rich in proteins – key nutrients for any growing organism. Watching the columns of wood ants returning to the nest heap often reveals individuals, or small gangs, manipulating dead insects back to the colony. Various statistics are available: 0.7–3.2kg of insect prey brought back to the nest each year; or a large colony

LEFT: Perhaps the most exalted meal ever enjoyed by British ants – *Formica rufa* foragers systematically deconstruct a dying female Purple Emperor *Apatura iris*.

BELOW: At the other end of the gastronomy spectrum, a wood ant *Formica rufa* tackles a fragment of cheese and onion crisp accidentally dropped onto the foraging trail.

bringing in up to 100,000 prey items per day. These comprise a whole range of groups from flies and moths to spiders, springtails, beetles, larvae, plant bugs, bark-lice and centipedes. Large items are ant-handled by two, up to a dozen, ants and the mathematics shows that a squad of ants can carry more if they work together to drag a large item than if they cut it up into the maximum load each could take on its own. It's all to do with turning the heavy object on a fulcrum, levering it from one static position to the next in a series of hoiks and hoists that would make a furniture removal company very happy if they could emulate it. In this case the whole is less effort than the sum of its parts.

Watching wood ants returning with their trophies is fascinating, but the range of their victims is not necessarily in proportion to the true fauna of the surrounding herbage. On the whole, most insects have a perfectly good defence against ant attack – they run, jump or fly away. Thus the species make-up of ant prey is probably skewed towards immature insects (flightless nymphs and larvae), diseased, dying or dead individuals, or those that cannot easily escape. I have often confiscated *Formica rufa* prey victims (for scientific purposes, obviously), including a large female *Rhyssa persuasoria*, a giant parasitic ichneumon wasp.

Despite the importance of invertebrate prey in their diet, it only accounts for a small proportion of food going into the *Formica* colony. A major component of the diet is honeydew. In one study a colony's intake was measured as 14.8–25.5kg/year, accounting for 78–92% of total food intake, and that was dry weight (Domisch *et al.* 2009).

Another estimate is for a large *Formica* nest to collect 500kg wet weight of honeydew per year (Zoebelein 1956). Even for large and apparently ferocious predatory ant species, honeydew is still the major food source.

Honeydew, tasting vaguely honey-sweet and looking like dew, is itself a by-product of an organism's search for nutrient protein. Sap-sucking Homoptera (mostly aphids, but also leafhoppers, mealybugs, scale insects etc.) penetrate leaves, stems, fruits or stalks with their beak-like mouthparts to drink down a plant's watery phloem contents. The phloem tubes transport the sugars created by photosynthesis (particularly the disaccharide sucrose), accounting for about 90–95% of the dry weight of the sap, most of which is water. But there is precious little in the way of proteins or their constituent amino acid building blocks – only 0.2–1.8% dry weight. Aphids absorb about half of the amino acids into their body structure, but in order to get enough protein for any reasonable growth, an aphid still has to suck up an awful lot of liquid, with the consequence, familiar to men of a certain age, that it very quickly needs to be got rid of at the other end. Throughput of sap can be quite astounding considering that these aphids are tiny insects; a large (5.0–6.0mm) specimen of the willow aphid *Tuberolachnus salignus* removes up to 40mm^3 of sap (that's very nearly twice its own body volume), containing about 4mg of sucrose, per day – the daily photosynthetic product of one or two willow leaves (Mittler 1958). Considering an aphid infestation in its thousands or

A *Lasius niger* worker stands guard over her aphid charges.

Myrmica ants busily tending a lively herd of black bean aphids.

hundreds of thousands on a plant, this is an exhausting herbivore load, even though no holes are physically chewed in leaves.

Aphid excrement is rather similar to the sugary liquid sap of its diet, but presents something of a danger to the aphid now that it is on the outside of the plant. It can mire the aphid in a sticky gloop; it can encourage mildew moulds which interfere with the aphid's food-plant; it can attract predators like wasps and ladybirds. Many aphids use acrobatic means to flick the sticky droplets away, and feeding on the underside of a leaf offers the possibility that gravity will take it away. This is the stuff that drips from trees and forms the sticky specks on the car windscreen when you park under a Sycamore or a linden tree in summer. Thankfully for the aphid, some ants have developed a taste for the honeydew and move in quickly to remove it, thus keeping the aphid colony clean and tidy. Actually, not all aphids are attended by ants, and not all ants attend aphids; but about 40% of world ant genera have evolved this mutualistic relationship, so even if not universal, it is nevertheless a highly important and widespread phenomenon.

The interaction is simple, well known and easily observed. An ant approaches an aphid from behind and palpates (tickles) it on the abdomen with its antennae. The aphid extrudes a liquid droplet of honeydew, which the ant drinks straight down. This action can easily be mimicked experimentally using a fine hair to tickle the aphid. The tip of the aphid's abdomen has a basket of bristles (the trophobiotic organ) so that the droplet is held suspended for the ant's convenience. There have been suggestions (first mooted by Kloft 1959) that the rear end of an aphid also resembles the head end of an ant – the back legs, often raised and waved about as the ant approaches, resembling antennae, and the two cornicles (also called siphunculi) resembling open jaws. Silhouettes purporting to show the similarities are widely illustrated in ant and aphid monographs and articles. Fifty years on, this is still a highly speculative suggestion awaiting experimental investigation. Ants also attend scale insects, mealybugs and plant-hopper nymphs for honeydew, but these other insects have behaviours and appearances that are entirely non-ant-head-like.

The cornicles stick out from the sixth abdominal tergite of the aphid. In many species these exude a sticky waxy defensive substance that readily gums up the jaws of potential predators such as ladybirds and their larvae, but in ant-tended aphids the cornicles are reduced, and this behaviour is not shown. Other defensive behaviours are also lacking in those aphid species associated with ant attendance – these include vigorous kicking with the back legs, and dropping off the plant altogether. One intriguing suggestion is that an evolutionarily precursor defence of kicking out wildly, raising the abdomen and ejecting the contents of the digestive tract, became ritualised when instead of attacking and eating the aphids, the ants simply drank down the sugary waste liquor but left the relieved aphid unharmed.

If sticky aphid honeydew dripping onto cars seems copious, this is nothing compared to the similar exudate sometimes produced by the mealybug *Trabutina mannipara*, which feeds on tamarisk sap around the Mediterranean and Middle East. Accumulated white powdery *Trabutina* honeydew is widely thought to be the origin of biblical manna-from-heaven stories (Bodenheimer 1947). At the height of the season a person could collect about 1kg per day, 'and it was like coriander seed, white; and the taste of it was like wafers made with honey' (Exodus 16: 31). The mealybug is regularly attended by *Polyrhachis lacteipennis* ants, but no mention is made of them in the Bible.

The milking of aphids for their honeydew has been a useful farming metaphor to describe their relationship with ants ever since Goedart & Lister (1685) imagined a conversation between stock animal and herdsman (Jones 1927). This metaphor can be developed into two epic extensions, which I shall now take and run with. First, just as milk becomes a commodity to be created, collected, moved, stored,

ANT AND APHIS. MILKING COW.

traded, and finally redistributed to a thirsty tea-drinking clientele, so too honeydew is the flexible and convenient food/energy currency of the ant colony. Kloft (1959) was partly inspired to make his silhouette guess because he had seen the meet-and-greet antennal testing and communication between two ant nest-mates when they came to exchange the liquid food. A forager ant full of honeydew is happy to be met by a fellow worker back at, or near, the nest and readily regurgitates the liquid into a droplet between her mandibles. The greeter drinks some or all of the honeydew offered and takes it off down into the colony, where the process is repeated in multiple mutual diminishing exchanges between nest-sisters. By this means, called trophallaxis, the harvested food is spread through all the members of the colony, whether they are workers in the field or engaged in other more constrained bureaucratic occupations elsewhere. Termed the social stomach (or social bucket), the collective guts of the colony members obtain nutrition (especially sugar energy) without the need for individual random foraging. The foragers ('farmers' or maybe 'milkmaids') bring back the honeydew, and it is then passed through various middlemen (technically perhaps middlewomen in the all-female ant colony) who

ABOVE: Demonstration of the milking analogy from the educational (= slightly condescending) *Nature's Teachings: Human Invention Anticipated by Nature* (Wood 1903).

BELOW: The idea that ants 'milk' aphids is now so well known that it even appeared on cigarette cards (1914). Sadly the nesting biology is a bit garbled, and the general emphasis is on destroying ants.

WILL'S'S CIGARETTES.

ANT "MILKING" APHIS.

A   B   ANTS & PART OF NEST.

eventually distribute it through the hard-working but home-bound populace, including, most importantly, the larval brood. Struggling on with this food/currency analogy, the bloated honeypot repletes we left hanging from the subterranean chamber in Chapter 2 are exactly analogous to food storage tankers buried deep in underground bunkers against some potential apocalyptic scenario – for ants, this is the sap- and honeydew-free famine of the dry season.

We can also stretch a second epic cow metaphor along the lines that the ants do not simply milk the aphids, they physically herd them about, and they also cull them like beef cattle. It has long been known that *Lasius neoniger* stores the eggs of the North American corn-root aphid *Anuaraphis maidiradicis* in its nests over winter and puts the nymphs out to graze on nearby roots the following spring (Forbes 1906). Ants will pick up and move live aphids from one place to another to improve the flow of honeydew, just as a farmer will change meadows to improve milk output. *Lasius fuliginosus* will move individuals of *Stomaphis quercus* on an oak tree to areas with the best honeydew production (Goidanich 1959), while *L. niger* will pick up and move late-instar *Pterocomma salicis* nymphs from one tree to another to establish new aphid herds (Collins & Leather 2002). Meanwhile they also frequently cull the aphids and eat them. Branching now down a veal analogy, a nest of *Lasius flavus* is reckoned to consume about 3,000 first-instar root aphids each day (Pontin 1978). This private stock of aphids is contained in the nest itself, but similar culling also occurs in aphid colonies above ground. The decision to milk or cull may be mediated by chemical markers placed on the aphids by the ants. In another study, individual foragers of *L. niger* frequently ate aphids that had not been previously tended by nest-mates, but 'milked' those that had previously been visited (Sakata 1994). A marker left by earlier workers is presumed to indicate ownership or productivity, just as cows are branded or ear-tagged on the ranch. Recently Leather (2017) showed that carefully washing maple aphids *Periphyllus testudinaceus* with dilute acetone caused the attending *L. fuliginosus* and *L. niger* ants to eat them more often rather than milk them. Swapping aphid-covered twigs between different ant colonies also causes more aphids to be eaten, suggesting that some colony-specific odour was involved in herd ownership recognition.

This is a metaphor that keeps on giving. In the Malaysian rainforest *Hypoclinea cuspidatus* farms the mealybug *Malaicoccus formicarii* (and other species) in true nomadic pastoral style by shifting the entire

ABOVE: Red ants, *Myrmica* species, exchanging food mouth-to-mouth by trophallaxis.

LEFT: Aphids are readily 'ant-handled' by their ant handlers. They are moved from leaf to leaf or plant to plant, although in this case the aphid is being carried into the soil, probably as meat for the young.

ranch about as the mealybugs exhaust the plants – the mass exodus includes the single queen, 10,000 workers, brood of 4,000 larvae and pupae, and upwards of 5,000 *Malaicoccus*. The bivouac may be 20m from the feeding sites and often moves about the forest to follow the best mealybug feeding areas.

Ants also control the physiology of their herded aphids. In a famous study of the common Black Bean Aphid *Aphis fabae*, El-Ziady & Kennedy (1956) showed that production of winged morphs in the aphid colony was reduced (by predation) by *L. niger*, who obviously wanted the aphids to stick around and produce honeydew food for longer. In the Japanese mugwort aphid *Macrosiphoniella yomogicola*, the attending *Lasius japonicus* ants artificially maintain their stock animals at a 2:1 ratio of green morphs to red (Watanabe *et al.* 2016). This is not simply an aesthetic imposition by the ants – it deeply affects the ecology of the entire aphid colony and its food-plant. It seems that although green aphids produce more honeydew, the red colour forms inhibit the mugwort from flowering. If the host-plant develops flowers, then goes to seed, this affects the sap flow and nutrient allocation across the entire plant and the aphid colony cannot survive. To prevent its food source dying out, the ants appear to maintain the optimum ratio of their charges, thus optimising honeydew production in the green morphs, but keeping enough less-efficient red morphs because of their beneficial environmental effects in keeping the nutrients in the host-plant in pre-flower full-sap flow. This is aphid farming on a par with landscape manipulation.

Despite predation, colony manipulation and physical interference by the ants, the aphids still benefit hugely from the ants' presence. As they give up their honeydew, and a few of their siblings, the rest of the aphids are being protected from other less caring predators and parasitoids, which the ants fiercely remove from the leaves. Many aphids (that Japanese *Macrosiphoniella yomigocola*, for example) and mealybugs (*Malaicoccus formicarii* with the Malaysian nomads *Hypoclinea*) can only survive if the ants are there to look after them, and colonies quickly perish if experimentally isolated from their herders. They become victims of outside predation, they suck their host-plants dry and starve, or they become contaminated by their own excess honeydew production and the resulting moulds. Here is a true mutualism evolved between the farmer and the farmed. It is an ancient model indeed: *Germaraphis* aphids have been found in amber from the late Eocene (40–50 MYA) associated with the ant *Prenolepis henschei* (Perkovsky 2011).

The most advanced, or at least the most complex, agricultural analogy must be saved for the leafcutter ants of Central and South America. About 200 species in 12 genera (*Atta* and *Acromyrmex* the best known) occur only in the New World (despite what the opening sequence of Disney's *The Lion King*, set very distinctly in sub-Saharan Africa, might suggest). Huge amounts of vegetable material are taken – an estimated 12–17% of leaf production (Cherrett 1986), up to 250kg (dry weight) per colony per year, reducing the cattle-carrying capacity of infested pasture by 10%. The leaf morsels are carted back to the nest and taken underground, but they are not eaten. Naturalist and explorer Henry Bates thought the leafcutters were collecting thatch material, like the wood ants he knew from Britain, but the truth was far stranger.

The ants use the cut leaves as a growth medium for their crops. The leaves are shredded into crumbs 1–2mm across, chewed and mixed with saliva and anal secretions, then inoculated with tufts of fungal mycelia. The mycelia grow at about 13μm/h and cover the surface of the leaf particles in about 24 hours. The ends of the microscopically hair-thin hyphae form tiny swollen bladders called gongylidia or staphyla, likened to kohlrabi heads by Möller (1893), who was one of the first naturalists to directly observe the ants eating them. The gongylidia are rich in glycogen (a polysaccharide of glucose), with some proteins, amino acids and lipids thrown in too. Actually it turns out that the adult ants get the majority of their energy input directly from plant sap, but as is typical with social insects, the larvae need a much higher protein food content, and the leafcutter grubs can only get this from the fungal kohlrabi harvest.

Despite more than 100 years of study, it is not altogether clear what fungus species are involved in the leafcutter nests. Millions of years of coevolution mean that the fungus is no longer a true wild species; to enhance the farming metaphor to the point of exhaustion, the

One of the large Mediterranean *Messor* harvester ants returning with a grass seed.

ants have effectively domesticated it. It no longer needs to produce spore-bearing fruiting bodies, and is entirely propagated from cuttings. One of the problems that this throws up for mycologists (and myrmecologists) is that fungal taxonomy and identification is normally based on the fruiting bodies that we commonly call mushrooms, toadstools or brackets. The only clue comes from some wine-red boletes (now named *Leucocoprinus gongylophorus*) found growing from abandoned *Acromyrmex* nests over 120 years ago.

Finally, to come full circle back to the seed-gatherers that inspired ancient Greek and biblical commentators, and our farming metaphor runs into the ground, quite literally. There is a myth that ants deliberately cultivate large-seed grasses near to their nests. But this is an accidental outcome of general seed collection. A germinating seed is useless for ant food, so most seeds brought back to the granaries are disabled by the ants biting off the radicle (the plant embryo). To stop nutritional deterioration the seeds also need to be kept dry; if they become wet because rain penetrates the nest, the ants carry the damp seeds outside to dry in the sun. Inevitably some seed radicles get missed and a few seeds are stimulated to germinate by the damp, in which case they are removed further and dumped around the periphery of the nest – this may have been the source of the cereal husbandry myth.

No wonder ant ecology has been studiously investigated and celebrated by fabulists, biblical propagandists and economists, using it to support all manner of human social models from socialism and the collective cooperation of the commune, to capitalist market forces, supply chains, production lines and militarism.

Growing – being fed enough for metamorphosis

The larva's job is to grow. Elsewhere in the insect world larvae are often free-living, as exemplified by lepidopterous caterpillars chewing their way through leaves. For the majority of these insect larvae, life is a lonely struggle, with high mortality in a dangerous world. Ants, by contrast, are cosseted. They are housed in the protective shelter of the nest, tended by nurse ants who guard them, feed them, move them, clean them and finally help them escape from the pupal cocoon after metamorphosis.

At its simplest, in 'primitive' predatory species like *Ponera*, larval nutrition comprises worker ants simply delivering prey items into the brood chambers and the larvae tucking in whenever they get hungry. Ponerines, with their narrow cylindrical abdomens, are less able to swell their bodies with liquid nutrients, so trophallaxis is uncommon. In the likewise 'primitive' Australasian *Amblyopone* and south-east Asian *Myopopone*, large prey items are left intact and the workers bring the larvae out of the nest for a picnic. Nests in these genera are, anyway, only crude shallow hollows under logs and stones, and the larvae, although not free-living, are corralled rather than housed. This movement – taking the grubs to the grub as it were – has also been seen in the European (and potentially British) *Aphaenogaster subterranea*.

Other semi-predatory or scavenging ants dismember and cut up the invertebrates they feed to their grubs. They also regurgitate the honeydew that has become part of the social bucket/stomach of the colony. Ant nutrition is relatively poorly understood compared to the similarly social honeybees, but given the importance of trophallaxis in those insects it seems certain that future research will discover at least some similarities and parallels in ants. One of the best-known (but perhaps least-understood) facts about honeybees is that workers secrete a proteinaceous substance from the postpharyngeal glands just inside the gullet, which they feed to the bee larvae. So highly nutritious is this substance that if it is diluted with nectar and honey the larvae become run-of-the-mill honeybee workers, but if the larvae are *only* fed the glandular secretion they enlarge to become well-fed queens. This is the royal jelly of overzealous health-food and alternative medicine devotees. So much of it is produced by the bees that the larger-than-average queen cells in the honeycomb are awash with the stuff – enough that it can be extracted and harvested by the beekeeper. No such queen-enhancing jelly substance has yet been found in ants (or indeed wasps), but there is evidence that secretions from similar postpharyngeal glands *are* passed from feeder workers to the fed grubs.

The cellular structure of these glands changes with the body composition of the ants, particularly in association with fat and lipid storage in the abdomen during winter; these stores are then metabolised and fed to the larvae in spring. Whether these glandular secretions contribute to queen, rather than worker, production, is still very circumstantial, but there have been reports that only young workers of the European *Formica polyctena* that had sufficient fat storage during the winter quiescence could rear new queens, and that this was directly related to the higher secretory activity of their

Nest of *Myrmica rubra* with larvae and pupae.

postpharyngeal glands. More on the queen/worker choices faced by growing larvae in the next chapter.

Having larvae in the nest is not a neutral add-on to the colony's routine; it changes the way foraging and feeding occurs for everyone. There is a constant conflict in any nest in that foraging ants make up only 10–20% of the colony workforce, but they must assess the nutritional needs for the colony as a whole and forage accordingly. The presence of larvae significantly skews nutritional requirements away from the norm, just as pregnancy changes the dietary drives and needs in a mammal. Adult ants, whether forager, soldier, builder or nurse, require carbohydrate energy to keep them going, but growing larvae (and egg-laying queens) need increased protein input. And here is the conflict – a high-protein low-carbohydrate diet seemingly shortens the lives of the ants. It appears that certain amino acids (notably methionine, serine, threonine and phenylalanine) are harmful to ants, possibly increasing toxic nitrogen waste products. To optimise their foraging efficacy, workers collecting the food must have some sort of feedback from the nest to tell them if they are doing it right. This is a promising area of ant research, and an excellent review is provided by Csata & Dussutour (2019).

Foragers do not work alone; as discussed later, they lay chemical trails along which they operate military-style supply lines. Despite

the fact that each ant has its own nutritional requirements, the laying of trails is not a simple algorithm based on food quantity, quality or availability – it is also affected by the overall starvation or satiation state of the entire colony, including whether or not protein-hungry larvae are present. The precise feedback mechanisms that are in play are still rather speculative, but an intuitively appealing chain-of-demand process has been proposed by Cassill & Tschinkel (1999), which I here paraphrase.

In the nest the larvae communicate their hunger to nurse ants, who in turn signal to foragers to go out and find food. When a forager returns and hands over, say, a high-protein morsel which is accepted by a nurse, this 'success' reinforces the forager to go and get more of the same. By laying and reinforcing a chemical trail to the food source the forager recruits others to continue the harvest of whatever the particular protein-rich items are. After a while, however, if the larvae are sated, they will stop eating the proteinaceous food offered to them by the nurses, who will be left not knowing what to do with it. Without an end-user for their current holdings, the nurse ants will then start to decline similar protein offerings brought in by the foragers, and these failures to hand over food will discourage foragers from seeking out repeat offerings. As demand for the high-protein foodstuffs wanes, those particular chemical trails will diminish and grow cold. Conversely, if the nest is carbohydrate-replete but protein-hungry, any foragers returning with sugary food will be left unable to offload it to their sisters, who will be eagerly seeking out nitrogen-rich tidbits in the food-exchange marketplace of meet-and-greet trophallaxis. So the social stomach registers hunger or satiation, and the foragers react according to their algorithmic understanding of supply and demand.

Left to right: early- and late-instar larvae of *Myrmica*, and four pupae, showing gradual coloration as the adult ant nears completion.

Meanwhile, back in the nest, the ant larvae are fed and generally tended by the broody ants. As with all growing insects the larvae moult every so often to shed their tight-fitting outer skin. A common analogy beloved of biologists employs the notion of immature arthropods outgrowing their tough suit of armour and having to shimmy out of it to allow the new one underneath to stretch and adopt to a larger body size before it firms up again. This works well when contemplating crabs or woodlice, but for the relatively soft and flexible ant larval skin perhaps a protective boiler-suit better fits the image of the hard-working blue-collar worker ant that the grub will eventually become. Each time it moults, the larva's resultant new flexible skin is larger and roomier than the last. The nurse ants are on hand to help with the removal of the old skin. In all British species there are usually four moults, but even a short hop over the English Channel finds *Crematogaster scutellaris*, which has three, and there are six in queens of some *Camponotus* species worldwide. As in many other insects, head-capsule size is a good proxy for the number of the larval stages, called instars. When the final larval instar is ready to pupate it can be moved to another part of the nest, away from the active larval feeding zone. Here it is bedded down with soil particles to form a cocoon.

Not all ant larvae spin a silk cocoon. The larvae of the Myrmicinae and Dolichoderinae form naked pupae, frequently encountered as pale pink ants-in-waiting if a *Myrmica* colony is uncovered under a log or stone. Ponerine and formicine larvae make a pale oval silk cocoon to protect the pupa inside. These are the ants' 'eggs' often found in garden colonies of *Lasius niger* and *L. flavus*, and once a common commodity in pet shops and used for fish food. When the adult ants emerge, the brood-chamber attendants are still on hand to help them escape their chrysalis and cocoon cases, and then to dispose of the empty containers. The adult ants are now ready to start work in the nest, and soon run off to find something to do.

Leaving the safety of the nest

Despite the frenetic ant caricature, a hatchling ant does not immediately take off at top speed to look for a picnic to invade. The nest is warm, dry and secure, the outside world is potentially wet and cold, barbarous and downright dangerous. Anyway, there is much to do at home before any ant needs to step outdoors. Food needs to be relayed through the colony by trophallaxis, tunnels need to be kept clear of debris, young grubs need to be looked after, fed and cleaned, and eggs need to be tended. A newly adult worker ant will usually spend several weeks in the dark nest focused solely on these domestic chores, before venturing out into the light. This transition from touch/scent navigation in the labyrinthine tunnels to the stark brightness of the outside world must come as quite a shock to the ant's sensory system. It has long been known that honeybees take faltering orientation flights around the hive before they set off on more adventurous nectaring missions. It turns out that bumblebees and wasps make similar explorations, so it should be no surprise that ants do likewise.

There is now a growing body of research examining how ants make their first expeditions outside, and although chemical scent detection is still important, it is obvious that many ants live in a visual world. Much of the work has been done on exotic long-legged predator/scavenger genera like *Cataglyphis*, *Myrmecia* and *Melophorus*, but the findings fit well against similar observations on, for example, our common red wood ant *Formica rufa*. There are distinct changes in the brain organisation after exposure of the visual system to light (Rössler 2019), and a ritualised or at least highly choreographed series of learning walks that each individual takes when it makes its first stumbling entrance out onto the world stage (recently reviewed by Zeil & Fleischmann 2019). These walks involve setting off from the nest entrance and making a series of data-gathering turns and twists – the ant is quite literally getting its bearings. There is a delightful glossary associated with these navigational trials as the ants work out 'home' and 'anti-home' vectors, by constant adjustment of positive and negative gaze direction combined with some sort of measurement of the Earth's magnetic field and analysis of the celestial compass produced by the polarised light pattern of the sun's rays in the sky. On their walks, the ants make full 360-degree pirouettes ('stop still and perform turns around their yaw axis') or walk in loops and small circles ('voltes'). During these turns, the gaze back to the home nest is fractionally longer than in other directions, indicating, perhaps, that the ants are scanning their surroundings, checking where the home nest is, and integrating the imagery of near and distant landmarks into something approaching a mental map of the area. Using an artificial arena, landmarks, magnetic field and sky polarisation can be experimentally manipulated to test the ants' abilities to navigate.

These learning walks are performed by 'naive' ants, and they always end back at the home nest. Mission accomplished. Successive strolls take the ants further out into the home territory, up to about 4m, at which point the learning walks are either too long and complicated for the researchers to follow, or the learner ants have passed their test and become subsumed into the general above-ground foraging population of the nest. It probably varies between individuals and between species, but seven learning walks over several days seems to be adequate training for your average grunt ant to graduate to full forager status.

Most ants negotiate the world, and orient themselves within it, using olfactory clues based on the trail scents they lay down as they go to and from the nest, or a visual map of the environment close to the nest, but those fastest-running of ants, the long-legged *Cataglyphis* in the deserts of Africa, use a method of dead-reckoning to calculate where they are in the world, and how to get back to the nest. When they leave on foraging missions, often across bare ground, they zigzag back and forth, circling and quartering the ground, looking for seeds, dead insects, or other scattered material that they can scavenge. When they find a suitable morsel, they do not retrace their steps, but instead set off in a more or less straight line back to the nest. In order to do this, they have somehow needed to compute where their apparently random labyrinthine wanderings have taken them. It's as if they have a map of where they are and how they got there. Now they calculate a navigational heading back to base and set off at a run, appearing to know the exact direction and precise distance required. It is thought that they do this by measuring the polarity of the sun's rays in the sky (using it as a solar compass), and adjusting for the sun's movements across the heavens, depending on how long they have been active. This is combined with a pedometer or step counter, to measure their way out on a forage trip, then calculate an equal or trigonometrically equivalent distance back to the nest.

Movement – walking, running, flying and standing still

Six legs is an optimum number for negotiating an uneven granular world, using an alternating tripod gait. Robot manufacturers should take note. This is a standard model of insect walking/running also shown in cockroaches, crickets, earwigs, beetles and non-hopping plant bugs. It works by having two legs on one side of the body and one on the other touching the substrate, supporting the body and moving it forwards, while the other three legs swing ahead ready for the next stride. This tripod walking movement is easily visible to the naked eye in large lumbering insects like green shieldbugs and *Carabus* ground beetles, but can also be visualised in ants using high-speed video telemetry; in a classic ant study by Zollikofer (1994a), individual footprints could be discerned as the insects ran across smoked glass – the smoke dusting being scratched by the tarsal claws and basal bristles. These marks could be identified as the triplicate sets of each tripod movement. Leg tripod geometry did not change with ant running speed, suggesting that the ants simply upped their pace when moving faster, rather than striding out. At the fastest speeds the footprint distances were further apart than the longest leg dimensions allowed, showing that the ants were 'trotting' – in other words they were completely airborne for the brief moments between tripod strides (Zollikofer 1994b). There was also evidence that, at the top speeds, the front legs lost touch with the ground, effectively making the ants partial quadrupeds as they ran harum-scarum. Tripod shape did change if they were bearing loads, as the ants shortened their stride to counterbalance the weight of a food particle held in the jaws (Zollikofer 1994c); when carrying weights the ants were effectively using their legs, not their backs. The tripods were also deformed as they negotiated curves – the outer leg arcs going round the curve remained the same as normal, but the inner leg arcs were shortened. In the future, this is how robots will negotiate bends.

Running speed in ants is fairly closely controlled by leg length and ambient temperature. In the British Isles various of the *Formica* wood ants are very quick, and *F. rufa* can deliver a reputable 5.06cm/s, easily comparable to the large Central American leafcutter *Atta colombica* (5.16cm/s) and the equally massive North American harvester *Verromessor pergandei* (4.27cm/s). Even the small *Linepithema humile* manages 3.09cm/s, but the leisurely *Solenopsis invicta* (the South American fire ant) strolls at a sedate 1.67cm/s (Hurlbert *et al.* 2008).

It's easy to appreciate this temperature-controlled speed in a personal experiment in the woods. In the oppressive heat of a sunny August day wood ants are far quicker running up the trouser leg as you wander over to admire the architectural details of their bustling nest. Only on a cool February morning is it possible to kneel at the nest side and sift through the thatch looking for ant-nest beetles while the dawdling ants shuffle about in slow motion in the cold.

Perhaps the best way to make personal observations is to look for uninterrupted forage trails up the trunks of trees and bushes. It is easily possible to time individual ants as they rush up or down between two markers – choose between a knot in the wood, a bud, a twig or an excrescence of bark and count how long it takes each ant to pass between them. After counting the times of, say, 10 or 20 ants, measure the distance between the markers and a quick bit of simple mathematics will calculate their average speed. The forage trails are when ants are at their busiest. These ants are on a mission, they are not simply idling about looking here and there for whatever they can find. They know where they are going and they go straight to it at top speed. More on these ant trails shortly.

Flying is not really that important to the everyday ant. Most adult insects fly, so the winglessness of worker ants is a distinctive and curious feature of their lifestyles. There are winged ants, of course – sexual males and females (queens) – but even in these individuals flight is a temporary expedient rather than a movement norm. The wings are a useful means of getting about to find each other when mating, but are not really used in everyday life. Most flying ants use their wings on only one day of their lives.

Ants can, however, make significant aerial movements, but most of this seems to be fairly passive, as they are wafted on the winds. Studies of flying ant 'migration' or dispersal are limited, but in one study of an invasive fire ant *Solenopsis saevissima* in the south-eastern United States, females were mostly found flying up to about 250m, and although the majority were estimated to move barely 1km, others occasionally made it 11–16km, up to a maximum of 32km (Markin *et al.* 1971). Males were recorded up to 300m, and in a classic study of aerial plankton over England, males of *Lasius niger* made up the vast majority of the 62 ants collected by a passive trawl net suspended from a helium weather balloon, although they were all found on just one July afternoon in 2002 when it was flying ant day in the area (Chapman *et al.* 2004).

Flying is fast, but is also a dangerous business; it exposes the ant to much quicker and more manoeuvrable predators, and to the wiles of the weather. On a sultry Saturday, 1 June 2002, my car was splattered by endless flying insects as I drove from Calais through Normandy on the family holiday to a gîte near Caen. Many of these were flying queen ants. That night a dramatic thunderstorm and lightning show lit up the sky and some of the insect remains were washed off the

OPPOSITE PAGE:
In early March, emerging wood ants *Formica rufa* congregate preferentially on the warm light-absorbing scorched timber of the stump in which they have made their nest.

On 3 June 2002 the sandy beaches of Trouville-sur-Mer looked streaked with silt lines, but these were actually countless millions of male *Lasius niger*, a few queens, and a scatter of other insects.

car windscreen by the torrential rain. But on Monday 3 June we visited the seaside at Trouville-sur-Mer to see the real results of the downpour. What at first I took to be rippling lines of black silt running across the pale sand turned out to be countless multitudes of dead *Lasius niger* males (Jones 2013). There were only a handful of queens (some still alive), but a rough calculation on the back of the proverbial envelope suggested 500 dead male ants per linear 10cm on each of at least five oscillating strandlines stretching as far as the eye could see, giving a conservative 25 million per kilometre of coastline. They had been part of a vast cloud of flying ants that had been drenched and deposited into the English Channel that night.

One aspect of ant movement which is very poorly understood at present is exactly how much time they spend standing still. Contrary to the stereotype notion that honeybees are in constant earnest busy activity, workers on the combs actually spend more time at rest than engaging in any hive-work (Kovac *et al.* 2011). It seems likely that ants, too, are not always as rushed off their feet as they may seem, and that they spend some (maybe a lot) of downtime sitting inert in the nest. It seems that only a small number of ants are ever out foraging at any given time. The majority appear to be standing still, grooming themselves or 'aimlessly' walking, according to one study that suggested workers of *Leptothorax acervorum* spend 78% of their time inactive (Allies 1984, quoted by Hölldobler & Wilson 1990). They only forage when absolutely necessary. A bit like teenagers. But with far fewer ant observation formicaria around than honeybee observation hives, we still do not know a great deal about their resting behaviour.

Orientation in an infinite universe

To a tiny insect, the world is an unimaginably huge space, stretching out immeasurably in all directions. Orienting itself in even a small plot of habitat full of plants and shrubs, boulders and puddles, barriers and impasses, is a task of monumental proportions. Visual acuity is not highly developed in the ant workers of most British species, and ant eyes are often very small, so they do not look where they are going in the same way humans might negotiate the route to school, to work, to the café, and home again. Instead, these ants rely on smelling their way about. In the tangled undergrowth and tight root-thatch, sight-lines mean nothing anyway to these tiny insects, but scents linger, and filter round objects; they can be detected in total darkness, and when taken with wind direction, they can tell heading and distance.

The biochemistry of these diverse pheromone scents is incredibly complex, part of the alchemical sorcery that makes ant communication and behaviour so fascinating. The evolutionary origins of these substances are still mysterious; many have antibacterial or antifungal activity and may have originated to counter infections. Across the Formicidae, different compounds may serve the same role in different species, or the same compound may have different functions. Unique molecules are produced by some species. Widespread chemicals may be manufactured in any of the several exocrine glands and have different effects depending on their relative concentrations, cocktail mixes of different substances and environmental influences. The study of ant pheromones is a scientific discipline in its own right, a study for many lifetimes. For non-biochemically-minded readers, some of the following few paragraphs may seem a bit tricky; you do not need to understand all the chemistry, just that there are lots of different chemical substances created by ants, that the ants can distinguish between them, and that they instinctively know which ones are important.

At its simplest, and best understood, ants leave scent trails as they depart the nest on foraging expeditions. They also lay trails from food sources back to the nest. Various species-specific substances are exuded from a variety of abdominal glands at or near the tail-tip. In most formicine ants, these pheromones originate in the hind gut; in myrmicines the venom gland is usually the source, or sometimes Dufour's gland on the underside of the gaster. Many hundreds of compounds have been isolated from ant glands, and identified using gas chromatography. Many others remain to be elucidated. One of the problems is that an ant's sense of smell is incredibly sensitive and the trail pheromones are laid down in unimaginably small quantities. The first ant-trail chemical to be identified was 4-methylpyrrole-2-carboxylate, in the Mexican/Texan leafcutter *Atta texana* (Tumlinson *et al.* 1971); it required 3.7kg of dried ants to isolate it, but when it was presented to the ants they showed a detection threshold of 80fg/cm, that's 8×10^{-14}g/cm, or 80 quadrillionths of a gram per centimetre of trail. One milligram of this substance, about the amount carried in the entire colony, was enough to lay a followable ant trail three times around the globe. Bonkers.

An interesting trail-gland exception occurs in the genus *Crematogaster*; at present none occurs in the British Isles, but they are widespread in Europe, occasionally get introduced in shipments of cut timber logs, and are potential colonists here in the near future. In this group the gaster is held up in the air, often reflexed over the rest of the body, away from the substrate, and the abdominal glands never make contact with the surface over which the ant is walking. Instead, glands in the hind tibiae have a duct to the tarsi, and the scent trail is laid with the back feet, the worker walking with a distinctive gait, holding its hind legs closer together than normal. In the African *C. castanea* the major element is the 12-carbon (R)-2-dodecanol, and in the Euro-Mediterranean *C. scutellaris* it is the 13-carbon 2-tridecanone.

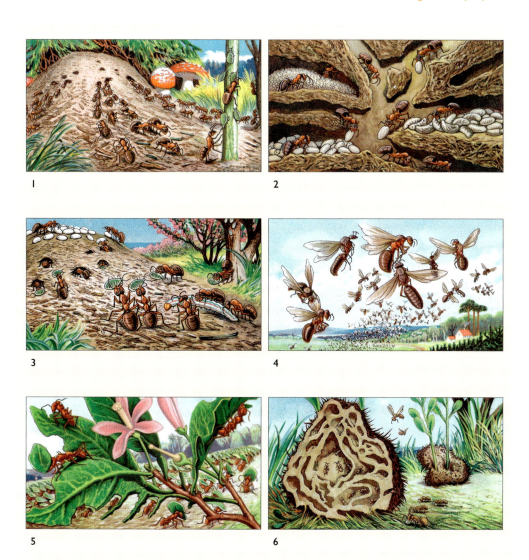

Collectable cards issued with Liebig (Oxo) stock/bouillon products (1932). Numbers **1** to **4** show wood ants foraging, nesting and in mating flight. The artist was not entirely sure how the leafcutters in number **5** carried their fragments, and seems to have been unaware that the *Myrmecodia* 'ant hotel' in number **6** is an epiphyte that grows high up in the trees.

These multitudinous volatile molecules vary from relatively simple 8-carbon methyl-heptanone $(C_8H_{16}O)$ exuded from the venom gland of the North American desert forager *Aphenogaster albisetosus*, through cyclic fatty acids like mellein $(C_{10}H_{10}O_3)$ deposited from the hind gut of *Lasius fuliginosus*, to 17-carbon terpenoids like faranal $C_{17}H_{30}$ from Dufour's gland on the underside of the pharaoh ant *Monomorium pharaonis*, and complex nitrogen heterocyclics like 2,5-dimethylpyrazine $C_6H_8N_2$ from the poison gland of the sting in *Tetramorium caespitum*. Many examples are given in a brief table by Hölldobler & Wilson (1990), and more extensively in reviews by Morgan (2008, 2009). I have raided these currently freely available articles wholesale for this section, so please refer to them and the very many references contained therein for even more detailed information.

At the technical end of biochemical study there is no easy appreciation of which chemicals do what, for which ant species. The diversity in molecular shape and size does not always seem to fit into any unified understanding of ant classification or relatedness. *Formica rufa* uses the fairly widespread insect substance, mellein (also described as (R)-3,4-dihydro-8-hydroxy-3-methyl-isocoumarin), as a trail pheromone, while *Lasius niger* uses the related (R)-3,4-dihydro-8-hydroxy-3,5,7-trimethyl-isocoumarin. Both of these compounds have been isolated from *Formica fusca*, but neither elicits any trail-following activity even though they have been shown to produce electrophysiological signals from isolated detached antennae under experimental electroantennogram (EAG) conditions. Likewise in *F. sanguinea*, two further EAG-active but trail-ignored isocoumarins have been found. Meanwhile, in the pan-Mediterranean dolichoderine *Tapinoma simrothi*, the pygidial gland (just behind the venom-gland and hind-gut openings) contains chemicals called iridoids, two of which (iridodial and iridomyrmecin) have a dual function – at low concentrations they act as long-lived trail markers along which the ants march sedately, but at high emission rates from a small focus they elicit alarm and many workers rush to the source with jaws open and abdomens raised threateningly.

Sometimes it is a cocktail mix of compounds that creates a followable trail. The present record for most complicated chemical recipe goes to the Asian ponerine *Leptogenys peuqueti*, which emits 14 low-molecular-weight alcohols and acetates from the venom gland. When these chemicals were synthesised in the laboratory they all showed mild trail activity, but only a mix of all of them gave a follower response equivalent to about 90% of natural venom-gland extract, suggesting that all of them were a necessary part of the signal line being followed by the ants. Sensitivity to a mix is even more extreme in Hong Kong army ants of the genus *Aenictus*. The major trail pheromone is methyl anthranilate, but the ants will only follow if 1% of the scent is the structurally similar methyl nicotinate. This tiny addition can be considered a primer pheromone, getting them ready for the trials of forage-trekking, because it can be given to the ants six hours before letting them onto the pure methyl anthranilate trail and they will then follow.

Despite all this complex chemistry, that ants follow a volatile liquid guide has long been known, and is easy to see in the simple experiment first demonstrated by Bonnet (1779) of drawing a finger over a trail where the ants are climbing up a wall. This causes the ants to temporarily lose the trail, and they will spend a few seconds searching around until they discover it again. At the start of Disney's *A Bug's Life* (1998) this is humorously played out as a leaf falls and interrupts the queue of harvesting ants; they have to be enticed to go around it until they rediscover the line on the other side. Other early experiments can also easily be replicated. As the ants walk it is possible to see them dragging their bottoms, and fine droplets of liquid are visible under the hand lens; these can be more easily visualised by dusting with fine powder as used by forensic fingerprint technologists, or the dark spores from a ripe puffball fungus if you have one to hand. Foragers will follow a similar trail laid with the gaster of a recently killed ant. In captive colonies, the floor over which a forage trail runs can be replaced and subsequent nest leavers, not knowing where to go, wander about aimlessly.

Although the scent is produced as a liquid, and is smeared onto the substrate behind the ant as it walks forwards, the best way to understand this trail is to imagine the volatile vapours arising from this line forming a corridor with a semi-ellipsoid cross-section in the air above it. This is the 'active space', an odour trail through which the following ants walk. The mix of various chemicals laid down together may become very important at this point. Many of the substances are highly volatile. The 8-carbon alcohol, methyl-heptanol, used by south-east Asian *Leptogenys diminuta*, is at the lighter end of ant alchemy, so this species also produces long-chain esters with 18-carbon oleate or stearate molecules tagged on to reduce its

Ant trails can be quite daunting to watch as thousands of individuals tramp past, obviously full of earnest diligence and focused activity. From the rather twee *Fairy Frisket; or Peeps at Insect Life* (Tucker 1889).

volatility and allowing it to remain longer on the trail. This is exactly what the carrier oil does in bottles of expensive perfume. Most ant chemical signals are based on a backbone of 5–20 carbon atoms, and with a molecular weight of 80–300 daltons.

Little is known about the precise mechanisms of olfaction in ants, or indeed in any animals. What seems probable is that smell receptors have an atomic-level three-dimensional structure onto which only certain shapes of odour molecules will bind to trigger a nerve response. Extrapolation from mammalian models suggests that, for simpler molecules, a receptor will bind to various odour chemicals but with differing affinities, and thus with different neural responses. Larger, more complex molecules have to fit very precisely into the receptor site in order to trigger a nerve. This is reflected in the fact that ant pheromone molecules may have several alternative three-dimensional atomic structures (even though they retain identical atomic chemical formulae) but receptors fit only particular structural versions. Historically these have been denoted as left- and right-handed isomers (S, *sinister* and R, *rectus*, for their left or right rotational movement of the direction of polarised light passing through solutions in the laboratory), but a non-superimposable mirror-image left or right glove analogy also works well. In the North American *Aphenogaster albisetosus*, the trail is made of the two isomers (S)- and (R)-4-methyl-3-heptanone, but only in a precise ratio of 8:2. In the Afrotropical *Crematogaster castanea*, laboratory-synthesised (R)- and (S)-2-dodecanol both equally led the ants to follow when laid down at 'high' concentrations of about 0.1ng per 32cm of trail, but below this the (R) form was 100 times more attractive than the (S) form.

That ants detect scent trails with the chemoreceptors on their antennae has also long been known, and was famously demonstrated by Hangartner (1967). He manipulated the ants' antennae before setting them off along the line. Normal ants tracked a regular slightly sinuous route, turning back to the line if one or other antenna left the narrow vapour corridor. After one antenna had been experimentally removed, the ants could still follow the trail, but with an exaggerated left-and-right zigzag course as they flailed, uncertain of their destination since 50% of their sensory apparatus had been deactivated. Ants whose antennae had been crossed over and glued into the 'wrong' orientation (straitjacket-arms-style) had great difficulty; a wildly staggering gait and back-tracking allowed them to only slowly follow the trail, probably helped by using light direction as an additional compass clue. It has also been suggested that antennae may have magnetosensory ability, with ultra-fine grains of various iron oxides being discovered in the pedicel of the antennae of a well-known South American migratory ant, *Pachycondyla marginata* (de Oliveira *et al.* 2009).

The pheromone trail is not permanent; otherwise the ants would eventually be overwhelmed and completely confused by weeks of chemical messages going in all directions, still hanging around in the environment. A typical trail, a mix of terpenoid farnesenes given out by the invasive fire ant *Solenopsis invicta*, seems to last about 100 seconds, then it is all but undetectable. However, part of the pheromone allure is to recruit other ants off to the food source. As these come and go, they reinforce the trail, adding their own glandular outputs and effectively raising the status of the track from single ant footpath to raging ant highway. Small colonies of the pharaoh ant *Monomorium pharaonis* are unable to forage when an experimental feeding dish is more than 50cm away; the trail pheromone is so volatile that it vanishes because it is not reinforced often enough. As the colony increases in size more ants can lay over-trails. The stronger this trail becomes, the more ants are recruited to scurry off to the food. In this way novel food sources can be quickly utilised and colony-wide effort can be directed to more productive food gathering. The trunk trails of multitudinous leafcutters and seed-harvesters can be in operation for many months.

Near the nest, each forage trail is usually a single pathway, easily observable as ants rush in opposite directions, but towards its end it may split into several sub-trails as, say, the ant foragers visit different groups of aphids clustered on different sprigs. Watching ants at these junctions shows that they do not randomly take left or right branches at a 50:50 ratio – they tend to diverge in direct proportion to the quantity of treasure that awaits them at the end of each branch. But how can they know whether to go one way to the larger target or another way to finish off the diminishing smaller prize? It's all down to the flow of ants before them, and how they reinforced, or not, the

A *Lasius niger* worker attending a small huddle of aphids.

trail. This is not, however, like a self-driving car robotically following road markings down the motorway, in which there is constant interaction between user and road surface. It turns out that the rules for ants leaving or following a trail seem to be:

1. On your way home, deposit a trail pheromone if you have fed, and are sated.
2. On your way out, choose a branch of a trail with a probability proportional to pheromone concentration.

Food sources ebb and flow. If they are on the increase, more ants returning down the route will be adding their own pheromone signal as they head back to the nest. If they get to the end of the trail and the food is drying up they will return and *not* emphasise the trail; thus the pheromone signals on this part of the network evaporate, and fewer ants are likely to take that side turning in the future. Some still do continue to take the lesser trail, and this is part of the variation in ant behaviour that might rediscover a new source of food, or a resurgence from that particular bunch of aphids a few hours or days later. Morgan (2009) and Czaczkes *et al.* (2015) give recent reviews of trail-pheromone chemistry, how trails are used by foraging ants, and how they regulate colony behaviour.

Ant trails like these are especially evident in the case of a larder invasion. It is easy to follow the ants back to the nest entrance, or to the gap under the door through which they have invaded the house. Here in the British Isles most ant trails are only a few metres long at most. But in tropical driver or army ants, the trails can extend for scores of metres, and the leafcutters, with their huge underground metropolises, often send out troops to a quarter of a kilometre.

Ant trails only really work, and are only clearly discoverable, in species with large numbers of individuals in the colony. In species with low colony numbers (perhaps just a few scores or hundreds), recruitment of workers to new food sources is not by mass chemical trailblazing but by a one-to-one personal tactile strategy called tandem running. This is exemplified in most species of *Ponera*, *Leptothorax* and *Temnothorax*. A successful scout returns to the nest and although it leaves a personal chemical trail on the ground it recruits a single new foraging partner by antennation (antennal touching) and regurgitation back in the nest. It then leaves the nest again, following its own scent-marking, but the follower, the tandem runner, remains in physical contact with the scout, using its antennae until they both reach the food source. If this escalates further, both fed workers now return to the nest and each recruits a new sidekick tandem runner until there may be dozens of pairs of ants running backwards and forwards. This physical/tactile communication shows similarities with the well-known circle or waggle recruitment dance of the honeybee and is very different to the heavily trod chemically mediated trails of larger ant colonies. No chemical trail scent has yet been identified for tandem-running species, but a volatile substance exuded from the scout's abdomen, and spread onto its hind legs by grooming, has been shown to induce the tandem runner to keep up on the outward leg of the journey. A tandem runner can also be experimentally induced to follow a severed gaster mounted on a pin.

Territory – it's worth fighting for

Despite the facile notion that ants bite and sting humans, usually those of us peaceably enjoying a summer picnic, it is mostly with each other that ants are at war. Ant ecology is dominated by competition between different ant groups, and that competition is more or less based on conflicts between competing colonies. This is easily seen from the spatial distribution of ant colonies across an apparently uniform habitat – they are generally further away from their neighbours than would be expected by random placement. When colonies die, or are removed, or are constrained within experimental barriers, neighbouring ants soon move in to take over the neglected foraging grounds. The balance of power can shift on a seasonal cycle. Small colonies concentrate on producing more workers, whereas large and well-established colonies gear up to producing the new sexual generation of winged males and females. Small colonies are thus able to insinuate their way into the territory mosaic. Again, it is easy to slip into epic metaphors, this time evoking colonial exploration, empire building, heroic warriors, deadly clashes, ritualised tournaments and raging battle scenes. This can be a useful analogy – though often not so much for the similarities with human conflicts as for the differences.

Whenever two ants meet, they test each other with their antennae (antennation), palpating one another. They are checking each other's smell, to see if they have met friend or foe. Each colony has its own unique chemical aroma, a complex cocktail of cuticular substances that are under a certain amount of genetic control, allowing sisters

Ants at war, *Formica cunicularia* and *F. rufa*, from *Episodes of Insect Life* by Budgen (1850). The original caption contains the statement: 'In the foreground is an instance, not uncommon in insects, of an individual retaining its vitality after the loss of its body.' This is typical of the tone of the time – bluntly educational, but also cosy and friendly.

from the same colony to recognise each other and distinguish interlopers. In human skirmishes, soldiers on the battlefield also need to be able to distinguish friend from foe – this might be in the form of uniforms, coat-of-arms banners, subtle regalia badges or secret passwords. A friendly greeting usually coincides with exchange of food and body fluids through trophallaxis, but strangers set to and fight each other. In ants, that is.

Of course, nothing is ever this simple, and there is a spectrum of peaceful–aggressive responses. Sometimes, if the 'alien' is of the same species, but from a nearby colony and perhaps sharing some recent genetic heritage, there is a casual tolerance. As the differences between two newly meeting ants become greater there is an escalation of possible responses – avoidance, jerking backwards and opening the jaws threateningly, nipping at each other, repeated biting, serious jaw-locking with pushing and shoving, and finally headlong biting and wrestling. As the aggression becomes more extreme, other ants are likely to get involved in the brawl.

It was once thought that ant queens originated some central nest scent, perhaps analogous to the swarm-controlling pheromones of queen honeybees, but this idea has now fallen into disfavour, even though queenless workers often merge with others from another queenless colony as if some vital recognition signal were missing. Incidentally, a queen honeybee's physical presence in the hive has long been appreciated by beekeepers, so much so that her absence (death by disease or old age) can be detected by subtle changes in the bees' behaviour, including the audible sound of the hive. In the archaic jargon of the apiarist a colony with a healthy queen is said to be queenright. I'm happy to borrow this excellent word for use in ant colonies. Sadly a queenless colony (bee or ant) has no such apt jargon term; queenwrong is just plain wrong. Some authors have used the term 'orphaned' to describe an ant colony bereft of its queen. I'm tempted to use a word (suggested to me on Twitter) borrowed from the history books – the monarchless interval after the execution of Charles I in 1649 and before the restoration of Charles II in 1660 – interregnum. Thoughts?

Current research seems to indicate that cuticular hydrocarbons (CHCs) on the body surface of ants are important in kin recognition; a useful review is given by Sturgis & Gordon (2012). These are small molecules, metabolically cheap to manufacture and readily dispersible, to be absorbed onto the epicuticle – the thin waxy outer layer of the ant's integument. In a recent study of harvester ants, gas chromatography identified many different hydrocarbon compounds in lipid extracts from sampled ants, and even though these all occurred in different colonies, they occurred in different concentrations (Wagner *et al.* 2000). Small glass blocks treated with nest-mate or foreigner hydrocarbons could readily be distinguished by the experimental ants, who threateningly flared their mandibles at the foreign glass but were more tolerant of nest-mate glass. Intuitively it seems highly unlikely that every ant colony could have a set of different chemicals to distinguish it from every other nest in the world, so having a colony-specific cocktail using the same palette of basic hydrocarbon molecular ingredients, but mixing them differently according to a unique colony recipe, is a simple solution that perfectly suits.

The source of these hydrocarbons is thought to be fat bodies inside the ant's abdomen, which exude the various molecules onto the cuticle surface. They mostly comprise saturated *n*-alkanes, methyl-branched alkanes

Bizarre Victorian Christmas card ostensibly showing 'the black ants being invaded by the red ants'. Contrary to the idea that season's greetings should be twee or kitsch, many of these late 19th-century messages were quirky verging on dark or sinister. There may have been some recognisable topical trope at the time, but the joke is now lost.

and unsaturated *n*-alkenes. These are relatively simple molecules, with 20–40 carbon atoms, similar to the paraffin wax of candles. They likely preceded colony behaviours and they (or very similar compounds) occur in the cuticles of many other insect groups where they help prevent desiccation and abrasion, promote water balance and create a barrier to invading microorganisms. The alkanes in particular have higher melting points and possibly evolved originally to aid waterproofing of the insect integument. Mixing of the various chemicals, however, alters melting and boiling points of the individual 'neat' substances, sometimes making them more volatile; the southern US *Camponotus floridianus* (frequently described as 'pugnacious') is able to detect nest-mates without antennation from about 10mm away, suggesting that ants are genuinely smelling one another, rather than tasting.

Constant licking and grooming of each other combined with trophallaxis food exchange helps homogenise the colony odour, spreading it among nest-mates throughout the nest. Aliens can nevertheless also acquire the colony scent. Experimental manipulation with workers from one colony being inserted into another shows that there is some tolerance if the 'intruders' have been able to absorb the colony's odour. This is possibly the mechanism by which the usurping queens of brood-parasite species (see page 185) can insinuate themselves into a nest, and is also how inquiline 'guests', mostly beetles (see page 254), move about many colonies with impunity.

It also seems that there is often some flexibility in the colony odour. If one ant meets another and 'tests' the intruder with its antennae, what exactly is it doing? It has always been assumed that each ant has some innate internalised olfactory template against which it is comparing those other ants that it comes across. But experimentation (and general observation) shows that there is rarely any exact overlap between template smell and the smell of any actual ant met on the ground. So is the decision to be friendly or aggressive mediated by the ant thinking 'How much do you smell like me?' or 'How different do you smell from me?' Is the ant measuring how much of the desirable familiar home cocktail is present? Or is it reacting to the presence of different 'foreign' smells? This recognition (or not) of nest-mates is often formalised in experimental studies into 'desirable-present' (D-present) and 'undesirable-absent' (U-absent) models as the ants compare the smell in front of them with the template they have acquired in their memory system. In disembodied electroantenna experiments, where neural signals from an amputated antenna can be measured electronically, it seems that the sensilla on the antennae are only responding to non-nest-mate CHCs; in other words they cannot smell themselves or their colony sisters, but can smell the alien aroma of a foreigner. There is, however, a grey area where nest-mates are sometimes apparently wrongly rejected, or non-nest-mates sometimes erroneously accepted, and testing these models, which might predict different acceptance thresholds, has preoccupied behavioural myrmecologists for decades.

It turns out that not all nest-mates in the same colony do in fact have identical hydrocarbon profiles; these profiles can change over time with the age of the individual, with the activity within or far from the nest that an ant mostly engages in, and with the season and other environmental effects. Changes in diet, especially in nitrogen-rich prey items, can also change the CHC profile. Conflicting *Formica rufa* workers from neighbouring nests are much more aggressive to each other earlier in the spring when the nests become active again after winter, possibly reflecting a need to better protect potential food sources, but also an example of straightforward opportunistic early-season cannibalism, since the vanquished ants are taken away, dismembered and eaten. During summer, workers from neighbouring *F. polyctena* colonies use the same foraging trails without aggression towards each other. It's as if they have realised that there is plenty of food to go around, and they have chilled out a bit. There is also a learning period – recently emerged callow adults are less aggressive to non-nest-mates, and unrelated callows can be placed with others to make mixed colonies. Perhaps they need to learn the template for the colony odour before they can start making satisfactory friend-or-foe judgements. They then use a combination of their own CHC odour and the combined nest profile scent to construct their internal acceptance or rejection criteria.

Out in the wider world, away from the close confines of the colony, kin recognition takes on the important role of battlefield recognition. Ants are quick to move in to exploit aphids or other ready food sources, and the pheromone trails through which they recruit extra help from fellow workers soon bring army-like numbers to bear on the harvest. Here is where sisterly recognition becomes important as territories are carved out, ownership of resources claimed, and the attentions of plunderers repulsed. The exclusion of non-colony foragers from the bountiful fields is an activity long practised by

humans throughout history, fighting over the local resources claimed by one tribe or nation state and defended from another – perhaps the oldest type of conflict in evolutionary terms. An excellent review of ant territoriality is given by Adams (2016).

In fact, territories are maintained by three means: fighting, guarding and avoidance. As with human conflicts, much of the territorial behaviour is ritualistic, and even in the largest of conflicts, many hundreds of *Atta* leafcutter ants may grapple with each other, but few die. I was tempted to use the term 'macho posturing' here, except of course all the ants are female, and 'feminine posturing' doesn't really sound right. I've struggled to find an appropriate word. Appeals to fellow etymological entomologists met with a certain amount of amused sarcasm and several daft suggestions such as wocho, facho and femacho posturing. The all-female Amazons could have given Amazonian, but this word comes with different connotations. Harridan, shrew, gorgon and harpy come with too much misogynist baggage. The closest anyone came was virago posturing. It's good, but I'm still looking.

Posturing aside, fighting to the death is common in large and aggressive species like *Formica* wood ants. These ants lend themselves to territory observation and experimentation. With the home nest so obvious and so large, and with a strong cohort of workers out on well-trodden foraging trails, their territories, expeditionary forces and

A *Myrmica* worker ant-handling another, dead, in its jaws. Whether one of its own colony, or a vanquished foe from a neighbouring nest, a dead ant is often ready protein for the grubs back home.

Dead red ants, *Myrmica* species – the aftermath of war.

combat zones are easy to see. In spring, when hordes of ants start to spread out in search of food, they frequently encounter neighbouring throngs, and the ants can die in their tens of thousands. The dangerousness of ant life has been calculated for the North American seed-harvesting *Pogonomyrmex californicus*: 0.06 ant deaths per worker-hour – mostly by fighting with neighbouring colonies (De Vita 1979). Whether because of massed losses, or learned avoidance, the sites of major *Formica* battles sometimes produce no-ant zones the next day, where ant numbers are very low and test-baits are unvisited. Ants are also less likely to organise foraging columns on the day after a fight, compared to days leading up to the battle. It's as if the ant colony is sullenly sulking in its hole, licking its wounds.

Unlike in human armies, no individual 'leader' ant has global information on the status of the colony's territory or on the availability or deployment prospects of ground forces; there is no centralised command headquarters from which instructions are given based on the intelligence received from observers in the field. No plans are drawn up or orders issued. Instead, any conflict, no matter how large, is merely an interrelated series of close-knit skirmishes, random jousts and personal duels. The outcome of these battles is usually determined by the numbers of combatants on each side.

The war on the ground – calculating who will win

Modelling of ant battles began with English polymath and engineer Frederick William Lanchester (1868–1946), who devised a series of differential equations (Lanchester's power laws) to try and understand how forces would be killed in idealised battle scenarios. In 1916 he did not start out to study the predator–prey interactions of wild animals, but instead the outcomes of the then very modern and still rather experimental conflicts of aerial warfare. For 'ancient' combat (hand-to-hand fighting with swords, clubs and spears), one soldier could realistically only fight one opponent at a time. Assuming some sort of equivalence in terms of weaponry, physical strength, armour and martial ability, it takes one soldier to kill one opponent. As the battle plays out, deaths in opposing armies more or less keep pace with each other, so that at the end of the battle the outcome is simply the difference between the larger army and the smaller, with the victors being the survivors of whichever started out as the larger force. Obviously training, nutrition, arms build-quality and morale will have effects, but we should perhaps blame *The Lord of the Rings*, Arthurian legends and the Icelandic sagas for promulgating any ideas of smaller and better-blessed forces overcoming the superior numbers of evil rabble hordes; in truth it is usually the larger army that prevails. Even the use of artillery, in the form of rather haphazard poorly aimed cannon fire, makes little difference. The rate of attrition depends on the number and density of available targets as well as

ABOVE: A grisly memento of an encounter in battle. The disembodied head of a *Lasius niger* worker attacker is now latched firmly to the stump of the leg it severed.

LEFT: Black (*Lasius*) versus red (*Myrmica*) ant wars. In this case the *Lasius* have the upper hand, or tarsus, as they spreadeagle their victim ready for dismemberment.

the number of cannons firing. Death and mayhem are still relatively equal on each side until the smaller force is eventually eliminated. Any advantage from the greater probability of a shot hitting the larger force is balanced by the greater number of shots coming from that larger force. This became *Lanchester's linear law*, since the strength of the army was directly proportional to the initial number of soldiers.

But with the advent of better guns, which could be better aimed, and especially with the appearance of highly mobile and manoeuvrable aerial guns (fighter planes), a soldier can now attack multiple targets, but also receive fire from multiple enemies. It now turns out that the fighting ability of a group of soldiers becomes proportional to the square of its number. Thus, doubling the number of the soldiers (or aeroplanes) now quadruples their effective fire-power. This became *Lanchester's square law*.

Adapting these calculations for conflicts between wild animals is useful, but problematic. There is general consensus that large ant colonies win over small colonies in major territorial conflicts. There is also agreement that summoning reinforcements quickly and efficiently can turn the tide of battle. Whether the linear law or square law is appropriate for ant warfare can, at least, be experimentally tested by the construction of mathematical models. If the square law predominates, this might explain why evolution has directed some ant species to concentrate on producing large numbers of small workers, rather than fewer large workers. The evolutionary decision of formicine ants to concentrate on firearm-style formic acid spraying might be explained by Lanchester's square advantage rather than the linear law which predominates in mandible-to-mandible close melee fighting. How does the appearance of Goliath-style major workers (not by coincidence called soldiers) on the battlefield influence mortality in opposing forces? Does ganging up, many workers spreadeagling then dismantling a victim, support the square law?

This is still a lively area of ant research. As might be expected in the chaos of inter-colony struggles, results gained by manipulating the theatres of ant war are never clear, but most evidence suggests that Lanchester's linear law comes closest to explaining battle outcomes. When it comes to ant armies, biggest is still best.

Gathering a large army into battle is often a result of just how quickly reinforcements can be recruited to the front line. Highly volatile alarm pheromones released into the air, or trailed along the substrate, alert and attract ants from many centimetres away to join the fray. Throughout the Formicidae, the mandibular glands, just inside the mouth, are known to produce many small volatile molecules, and substances like ethyl-ketones, 3-alkanols, terpene aldehydes, alcohols, 3-octanone and 3-octanol are usually identified as alarm substances, alerting and attracting nest-mates to come and investigate the source.

This chemical signalling can soon escalate to chemical warfare. A typical small ant may carry 40µg of an alarm/panic/aggressive warning pheromone, but the large North American *Formica subintegra* has a huge Dufour gland in its tail which can eject 700µg of what are usually described as 'propaganda' substances. *F. subintegra* is a slave-making species (see page 190), and during its raids on other ant species this spray causes panic, disorientation and chaos in the target nest.

Organised fighting is the most conspicuous way that ants partition the available landscape, but just as in human border disputes, not everything escalates to inevitable bloody destruction. Territories are also maintained by more subtle behaviours. Border guards patrolling around the perimeter can intercept and neutralise foraging scouts so that they never return to their home nest to recruit extra forces. If the guards are experimentally removed, other ant colonies move in and take over (Lescano *et al.* 2014). Ants also self-restrain by avoiding entering foreign territory. They recognise local scent-mark cues laid down by the home colony and accept them as no-entry signs – keep out, trespassers will be killed and eaten. In tropical arboreal ant species, large battles can be sparked by experimentally moving nests and bivouacs, forcing the ant colonies to intermix, but in the normal course of events these escalated conflicts are rare and a peaceful, if uneasy, territoriality is maintained by guard patrols. Workers experimentally placed into a neighbouring territory become agitated and cautious and retreat to their own side of the boundary as quickly as possible, avoiding confrontations with the home-team ants and making submissive supplicatory signals if they are challenged. Conversely, ants within their own territory are more aggressive to intruders, for they know the advantage of being on familiar ground, within easy signalling distance of armed back-up. Indeed, ants adopt a mob mentality by being more aggressive if in the company of about 20 nest-mates than if they are alone. All in all, there is quite a significant overlap between ant and human aggression.

The rise of the colony

W hat ants lack in size they more than make up for in numbers. And these numbers are housed in that most ant-like of creations – the nest. There is a tendency to anthropomorphise ant nests, and liken them to human communities – towns, cities, nations. I did this myself earlier. But there is an important distinction often lost in this comparison – the motivation of the occupants. Where humans (supposedly) expand and build to increase and improve the lot of society, to promote civilisation, to create culture, and make the world a better place for us all, ants are driven to build cities purely to produce the next generation of sexual queens and males. The massed citizenry of worker ants is simply a biological by-product, a necessary (but ultimately disposable) stage in the accumulation of nutrient biomass for, and in defence of, that royal blood line. I'm sure there's an ugly capitalism metaphor here somewhere.

This worker disposability is a feature of all the truly social insects – ants, honeybees, bumbles, yellow-jacket wasps, termites. With their worker castes and ruling monarchic societal structures they are termed 'eusocial', as distinct from the quasi-social nesting aggregations of, say, a group of leafcutter bees all nesting in nearby tunnels in a dead tree trunk, or the gregarious feeding habits of processionary moth caterpillars. Eusocial insects have all been celebrated for their industry and architecture, and for their complex societies showing apparent municipal planning, civil defence, care of the young, cooperative foraging and equitable food distribution. In times past these insects were held up as natural examples of a godly social structure where food and shelter, community and safety, calm and stability, were the just rewards of a proletariat worker class overseen by a ruling aristocratic elite. The socialist in me shudders.

OPPOSITE PAGE:
The colony exists with one aim – to create the next generation of flying ants.

175

Today biologists view these social insects slightly differently, but the complexities of their colonial lives are still fascinating. What makes ants so special is that *all* ant species are eusocial, from the most 'primitive' ponerines, with nests of a few dozen individuals, to the megalopolises of the leafcutters.

Nesting by default

The textbook nesting strategy of ants is very well exemplified by our commonest species, the black pavement ant *Lasius niger*. It's a good place to start – in July or August. As the skies fill with flying ants, queens and males pair off and eventually drift to the ground. They are unlikely to fly far, perhaps just a few hundred metres, since their glycogen energy reserves are depleted soon after take-off. Laboratory experiments show that ant 'nuptial' flights can be as short as 1 minute for some species. After uncoupling, a male's purpose is past, and he will not live long – just hours probably. It is unlikely that he will mate a second time. The inseminated queen, on the other hand, may reach 30 years, if she's very, very lucky.

First, though, she will have to find a safe place to lay her eggs, and here begins her long-game strategy. She lands and immediately takes shelter. Raking her body with her middle and hind legs, she breaks off her wings at natural shear points – the basal dehiscent sutures.

Attached to the much larger female, the male black ant *Lasius niger* becomes a mere passenger and goes wherever she goes.

She need never fly again, and the wings would just get in her way for the rest of her wholly underground life. Judging from my own observations, the large front wings detach before the smaller hind wings. Male ant wings do not break off so easily – there is no need, he has no subterranean career to look forward to.

Taking up residence in a tiny crevice in the soil, the newly mated queen excavates a small hidey-hole and seals the entrance with soil particles. Down here she lays her first eggs and stands guard over them in the darkness. When the grubs hatch she feeds them using more eggs, not viable ones this time, but infertile 'trophic' eggs. Typically these are flaccid, or differently shaped from reproductive eggs, and they usually lack DNA. In the South American leafcutter *Atta sexdens*, the oocytes fuse and the queen delivers a trophic omelette. Yum.

Effectively the queen is converting her body, metabolising the thoracic wing muscles she will never use again, and the fat stores in her abdomen (something like 40–60% of her body weight), into convenient nutritional packets for immediate consumption. Her first brood of 10–20 fertile eggs hatches after 6–10 weeks and the larvae reach maturity and emerge as adult workers in about another 4–8 weeks. They are often very small (called nanitics or minims) compared to workers reared later in the colony's life. They are also very timid, not at all aggressive, and will avoid any confrontation; this is a prudent choice, since for the fledgling colony to continue, the survival of these first few workers is paramount. They leave the nest chamber to find honeydew or whatever scavenged nutrients they can get hold of. At first they feed only the queen, who continues to produce trophic eggs, but eventually there are enough workers to source nutrition for the queen, the brood, and each other. The queen, meanwhile, continues to lay more brood eggs, and from now on, for the rest of her days, this will be her sole job.

The *Lasius* queen tactic of hiding in a secret soil recess and using trophic eggs to bring on the first batches of workers is termed claustral nesting, from the Latin *claustrum*, meaning lock or enclosed place (which also gives us claustrophobia, and, by a slightly more circuitous route, the word cloister). Claustral nest foundation is considered to be slightly more evolved than the non-claustral nesting used by, say, *Ponera*, which is regarded as one of the 'primitive' ants. *Lasius* queens are much larger than the workers (about 30–40 times as heavy), an adaptation which gives them the necessary body reserves so that they can hide away without feeding themselves, and use trophic eggs for a month or two to get the colony going. A *Ponera* queen is roughly the same size as her workers and has limited body reserves. Rather than retiring to a claustral chamber, she must constantly leave her eggs and go off hunting and foraging, returning to the nest every so often to check the brood and feed the hatchling grubs. This is thought to reflect an evolution from predatory wasp-like ancestors – and it is the nesting behaviour of all modern wasps, from solitary ground-nesting species to the social yellow-jackets. In fact, various non-British ponerine species show a series of graded steps from non-claustral, through semi-claustral (where some body reserves are metabolised but bolstered by occasional foraging), to fully claustral. Founding a new nest in the secret security of the claustral chamber is intuitively a more advantageous strategy – foraging is a dangerous business and workers suffer their highest mortality on trips out from the nest. Queens forced to leave the brood cloister and search above ground for food are in equal mortal peril.

Once the first workers of the new colony are able to forage, there follows a long period of colony growth and expansion, often termed the ergonomic stage. New subterranean galleries are excavated and the queen lays more eggs in them. Within a few months the population of the nest is increasing almost exponentially, but still it is only workers that are being reared. At some point, though, a physiological trigger initiates the next phase of the colony life cycle and batches of eggs are produced that are destined to become new queens and males. In fact the colony life cycle can be distilled down to the idea that energy investment in the nest is concentrated on multiplying up worker numbers until such time as it becomes profitable to convert some of the net yield into the development of queens and males (other financial and economic metaphors are available).

Exactly what triggers this change to generating sexual reproductives differs from ant species to species, and from colony to colony. Intuitively (as is relatively well known in honeybees) this is likely to be under hormonal or pheromonal control, reflecting colony size, ant biomass, nest age, inflow of nutrients, worker activity and readiness, as well as having some seasonal input in terms of daily temperature. In honeybees the queen secretes a pheromone from her tarsi which gets smeared across the honeycomb as she walks, and this footprint pheromone inhibits workers from making the special large cells in which new queens will develop. There is good evidence that the presence of an ant queen can initially inhibit a new sexual generation. In experimental nests of the fire ant *Solenopsis invicta*, colonies without a queen, or with only one queen, start to produce new queen and male larvae within four days, suggesting dilution of some chemical in the nest that normally inhibits production of a new sexual generation, but colonies with multiple queens do not. No specific chemical compound has been identified, though.

The question of which processes control the larval developmental switch from worker body form to queen is one that has occupied the wit and wisdom of myrmecologists for decades, and the answer is almost certain to lie in a combination of both hormones and nutrition. A useful review is given by Trible & Kronauer (2017). In social wasps there is good evidence that more nitrogen-rich prey produces queens rather than workers. This is achieved by the foraging wasp workers offering their putative queen grubs predatory insects like aphid-feeding hoverflies, hunting solitary wasps, or carrion-feeding blow flies rather than plant-eating caterpillars, froghoppers or greenfly. Experimental manipulation of ant diets has not provided much clear evidence of whether a similar dietary switch triggers queen development. Nevertheless, the switching or decision point at which a larva commits to becoming a queen or a worker has been measured. In *Myrmica ruginodis* the queen/worker bifurcation point in the developmental process occurs very late in larvahood – about a week before the larva is fully grown; until that point anything can happen. This rather suggests that nutritional factors are not that important. In *Formica polyctena*, on the other hand, the final endpoint decision has already been made within 72 hours of the larva hatching.

In honeybees the important nutritional contribution of royal jelly secretions is well known, but nothing like this has been found in ants. However a developing female ant arrives at its final worker or queen destination, though, it is not down to complete chance. The change to rearing queens and males, when it comes, is not some sudden irrevocable alteration in ant behaviour, but a subtle adjustment to existing behaviours. Eggs are constantly being laid in the nest, but what happens next is that some of the female eggs are destined to become queens, and male eggs also begin to be laid. This usually happens during the second or later years in a successful colony's existence.

Young queens, during the first year of colony life, are more likely to concentrate on producing only workers, to build up the size and strength of the nest. In temperate northern Europe, optimum growth conditions (usually good food availability and temperatures

Cocoons in a *Lasius niger* nest. The two sizes show worker and queen pupae ready to hatch.

above 20°C) first produce workers. Conversely, winter chilling leads to synchronised larval development, first by slowing or stopping metabolism (diapause), then by reawakening it in spring, and this leads to more queens developing, with growth trajectories aligned to arrive at bodily completion around the same time for that all-important synchronised nuptial flight emergence.

The mating game – nuptial flights

The end product of colony growth, though not necessarily the end of an ants' nest, is the production of the next generation of sexual males and females. These often appear on the well-known flying ant days when countless millions of queens and males take to the air in an orgy of synchronised coitus. Queens and males often reach larval maturity at different times, even if they have gone through diapause – being slowed by winter cold and reawakened in spring. But they do not simply climb out of the nest and fly off as soon as they are ready. Instead, they emerge from their cocoons and wait inside the nest chambers until conditions are exactly right. In summer it is quite common to find winged sexuals in an ant nest, even though that particular species may not be ready to take to the air for several days – they are, to borrow a term from the aviation industry, flight-ready.

Male *Lasius niger* ants readying themselves for a massed aerial nuptial flight. As with many Hymenoptera, males appear just before the females.

Quite what 'exactly right' conditions might be varies from species to species and from site to site, and precisely measured temperature or meteorological details seem broad and not very specific when calibrated against mass flying ant emergences. For example, looking at *Lasius niger*, Boomsma & Leusink (1981) recorded air temperatures of 15.9–23.8°C, relative humidity of 42–70%, global radiation (a measure of direct and refracted sunlight) 133.6–262.9J/cm^2/h; these are all quite broad ranges and could be typical of almost any afternoon in July or August when this species generally takes to the wing. Similar broad parameter ranges were found for mating flights of *L. flavus*, *Myrmica rubra* and *M. scabrinodis*.

The idea of a nationally coordinated flying ant day is also an urban myth. Although there is some synchronicity between nests in particular small local areas, there is no mass emergence across any large swathe of the country on any particular date, and ant clouds can fly almost any time from June to September, varying year-on-year. Having said this, vast clouds do sometimes appear as local synchronisations in small local areas become super-synchronised across larger geographic regions. As mentioned in Chapter 5, I had seen the end result, dead, drowned, on the beaches of Trouville-sur-Mer in 2002, and in mid-2020 a massive cloud of ants registering on meteorological radar around the coast of Sussex and Kent gave rise to some ill-informed and less than helpful swarm/invasion/end-of-civilisation headlines in the tabloid press. Hart *et al.* (2018) found that all days with a mean temperature above 25°C and wind speed less than 6.3m/s had ants flying somewhere in the country. They did, however, find that ants were pretty good weather forecasters. On almost all flight days, the weather (temperature and windspeed) was better than on the day before, suggesting that the alates were waiting for a subtle but important meteorological improvement. Mating flights often happen the day after rain, when the air is humid and the soil is moist, and thus easier for the newly impregnated claustrally minded queens to excavate.

Synchronisation in the air offers two important advantages to flying ants – safety in numbers and a party atmosphere to get to know each other. Hiding in plain sight is a good defence for a small vulnerable insect struggling with blustery winds. When the swifts and swallows come careening through the air, the chances are that some other ant will end up as lunch – if there are enough flying ants up there at any one time the odds are they will miss you and gollop down one of your brothers or sisters instead. Meanwhile, if all males and females from a wide area appear at the same time it helps ensure easy and convenient mating opportunities between nests, preventing too much inbreeding. To a tiny insect the outdoor world represents an almost infinite void. Locating one another is a journey into the unknown. Some insects get round this by congregating at a mutually attractive foodstuff (dung flies around a cowpat, for instance), signalling via an airborne chemical scent (many moths, especially at night), rendezvousing at flowers (hoverflies, soldier beetles) or prominent landscape features like trees (bumblebees, wasps). But if by sheer numbers the air is thick with potential mates it makes it all the easier.

Ant mating is quick. There seems to be little in the way of any aerial courtship, and males simply grapple the females. They then fly in tandem – the larger female usually supporting the weight of the diminutive male. When they land the female sets the agenda by walking where she wants to go and the little male is carried with her. Eventually they decouple, often causing damage to the male genitalia; he will die shortly anyway, having successfully completed his sole biological task.

Queens (and one male) of *Lasius niger* ready to take to the air.

Queen and male *Lasius niger*, showing the large size disparity. Once coupled it is the queen who determines where the journey ends.

Though nuptial clouds appear on warm summer days, this is not a simple response to temperature; queens also need an ambient temperature around 20°C to develop. At the northern edges of its range in Scotland, *Myrmica rubra* nesting in shaded woodland is able to build nests that give rise to workers, but the 20°C threshold is never breached, and so queens are never produced. Indeed, few ants anywhere in the world function well below 20°C, and none seemingly below 10°C. The diapause-induced synchronicity is a combination of warm summer and autumn weather first, then the cool of winter. Larval nutrition and ultimate larval size are also important, but only in conjunction with other influences; in *Myrmica ruginodis*, a mature larva reaching 3.5mg after winter chill will become a queen, but without that chill it will become a worker no matter what its size.

One surprising observation of nest growth is the fact that worker ants are not as seemingly expendable as, say, honeybee workers. At the height of busy summer, honeybee foragers live only a few days or weeks (they work themselves to death) and their corpses often litter the ground near the hive entrance, but worker *Lasius niger* ants live much longer – average life expectancy of 310–430 days, but up to 1,129 days recorded (Kramer *et al.* 2016). Nevertheless, in a large colony mortality is high, and many ants die every day. Just like honeybees, the ants do not allow their dead nest-mates to moulder in corners (that would be a disease-fuelled health risk), and they indulge in the hygienic removal of bodies called necrophoresis. This has long been observed, but contrary to some rather soppy

BURYING-ANTS.   SAVAGE FUNERAL.

interpretations there is no evidence of ant cemeteries, or intentional burial. Instead, the corpses are unceremoniously dumped on refuse piles near the nest, in 'kitchen middens' along with uneaten prey items, or in special chambers set aside to receive assorted nest rubbish. The living ants are alerted to a death by a small range of simple decay chemicals, notably oleic acid. In a test experiment, this substance can be extracted from ant corpses using acetone; living ants doused with the oleic extract are picked up by their sisters and thrown (quite literally) out of the nest. If they return without cleaning themselves properly they are cast out again.

There had long been a notion that ants deliberately, if primitively, buried their dead, hence this rather culturally insensitive illustration from *Nature's Teachings: Human Invention Anticipated by Nature* (Wood 1903).

Building on the work of others

Making a nest is a difficult, dangerous and labour-intensive process, so moving in to someone else's is a strategy well worth considering. This is termed social parasitism, or nest parasitism, and ants have evolved several different ways to borrow, invade, take over, or otherwise usurp the work of an existing colony. In some bee and wasp species this behaviour has been called cuckoo parasitism, echoing the interloping behaviour of the bird well known for laying its eggs in the nests of others. Ant parasitism is probably quite common, although only about 230 ant species (from a total world fauna of about 12,500) are firmly recorded as nest parasites. A census of the well-known Swiss ant fauna showed that almost a third of the country's 110 species were nest parasites (Kutter 1969), and at least 13 (26%) of our nominally 50 native British ant species are parasitic in this way. There are possibly many more socially parasitic species waiting to be discovered among the poorly known tropical ant faunas. Parasites in any group of organisms are usually scarcer and more difficult to find than their hosts anyway, and this is also true of socially parasitic ant species. The diversity of different ant colony structures has meant that social parasitism has had plenty of opportunity to come up with quite a few variations on the theme. Broadly, though, they are grouped into four different types.

Guest ants

Xenobiosis, from the Greek ζενο (*xeno*, guest) and βιος (*bios*, life), is just that – one species of ant living the life of a guest in the nest of another. The small, rather slim and shiny *Formicoxenus nitidulus* only occurs in the large domed heaps of its hosts – in the British Isles the wood ants *Formica rufa*, *F. aquilonia* and *F. lugubris*. It makes its own small nests inside hollow stems and twigs that have become incorporated into the fabric of the host structure. *Formicoxenus* only occurs in association with its hosts and cannot survive or make independent nests without them. Nevertheless it has its own workers, which rear the *Formicoxenus* brood, and eventually the next generation of *Formicoxenus* queens and males. It appears to solicit food by trophallaxis from its hosts, and although it will feed on *Leptothorax* larvae if offered them in experiments, it has never been observed to attack *Formica* young under artificial or natural conditions. It also gets shelter from the weather and protection from potential predators by taking up residence in the well-fortified wood ant fastness. Even so, the exact cost to the host is not necessarily very onerous, since *Formicoxenus* colonies are very small (up to about 100 workers) in a host nest of maybe a third of a million wood ants. Arguably the parasite is leading a reclusive and secretive life in a quiet corner and not doing too much harm – a bit like the silverfish living under my cooker.

Xenobiosis is not common across the world, and is only recorded in three genera – *Formicoxenus*, *Polyrhachis* and *Megalomyrmex*. In all cases host and parasite are only distantly related in evolutionary terms and usually very different in size. *Formicoxenus* is tiny compared to *Formica*, and its nest tunnels, extended through a hollow stem of dead bracken or small twig, are too narrow for the *Formica* workers to get in. The parasites are ignored, or tolerated, and are thought to acquire the kin-recognising cuticular hydrocarbons from their hosts within the first

Formicoxenus nitidulus, living up to its English name 'shining guest ant' – the reflection of the waving photographer can be seen on its body.

few days after emerging as adults. They also have their own repulsive cuticular smell, and if one is picked up in the jaws of a *Formica* worker, it is dropped immediately.

Temporary social parasitism

Though temporary, this is not just short-term squatting in someone else's nest. It may not last long, but it still ends with the destruction of the host ants and a complete takeover by the parasite. It begins with a mated queen of the parasite entering the already well-established colony of the host species. She must first acquire the odour of the colony, which she does by skulking in quiet corners for a few days. Queens of *Lasius umbratus* have been recorded attacking and chewing workers from the *L. niger, L. flavus* or *L. brunneus* nests into which they were insinuating themselves, whereby they presumably absorb some of the nest-specific cuticular hydrocarbons. The invader queen then seeks out the host queen and kills her, often by grasping the neck and throttling or decapitating her. In the non-British *Leptothorax goesswaldi,* an invading queen attacks and kills the host queen of *L. acervorum* by gradually biting off her antennae. What a horrible way to go. The new queen is quickly adopted by the host workers, who continue to go about their normal foraging, defensive and brood-tending duties. They seem unaware that their monarch has been replaced and that the new eggs, then larvae which they are rearing in the brood chambers, are of a different species.

When the first parasite workers emerge as adults they mingle with the host species and the nest functions as a mixed colony for some weeks. Eventually, though, since they are not being replaced, the host workers die off by attrition, and the nest becomes a pure society of the parasitic species. *Formica lugubris*, *F. exsecta* and *F. pratensis* are temporary social parasites in the nests of *F. fusca* and/or *F. lemani*, and there are suggestions that many other *Formica* species, including the common *F. rufa*, invade already established nests of closely related or indeed their own species. Sometimes they form dual or multi-queen (polygynous) colonies from which a queen and a cohort of workers eventually bud off to found another separate nest nearby.

The parasites do not necessarily just waltz in and take over; there is evidence that the host ants fight back – or at least their larvae do. There is always a certain amount of cannibalism among the ant brood, and Pulliainen *et al.* (2019) showed that host *Formica fusca* ant grubs preferentially devoured the eggs of the *F. exsecta* parasites laid alongside them. This may seem like rebellion, but if the host queen is dead, the rebels are merely putting off the inevitable.

Lasius fuliginosus is a temporary social hyperparasite. A queen invades and eventually takes over a nest of *L. umbratus*, which is itself a temporary social parasite in nests of other *Lasius* species. This distinctive shining black ant is widespread in England and Wales, but is never common. Its highly complex colony creation strategy makes successful establishment an order of magnitude more difficult with every layer of parasitism involved. I got very excited when I found it in Nunhead Cemetery, south-east London, in 2002 near nests of *L. brunneus*, although I did not find *L. umbratus*. As far as I know, this is the most inner urban London record on file. *Lasius brunneus* still occurs regularly, but I can no longer find *L. fuliginosus* there. If that colony failed, it may take some time for all the planets to realign, such that colonisation can occur again. According to Donisthorpe (1927a), *L. fuliginosus* nest cartons, made of papery chewed wood particles, are said to have a characteristic pungent aroma, although I can't say I've noticed – maybe I didn't get my nose close enough. This ant's head is supposed to have a violent odour when crushed – at least according to those who have done the crushing. The alitrunk and gaster are also, apparently, not entirely inodorous. Perhaps there is scope for a series of crushed-ant smelling experiments.

Permanent parasitism

Sometimes called inquilinism (from Latin *inquilinus*, a temporary resident), permanent parasitism occurs widely across the world, and is the most common form of social parasitism. Instead of killing the host queen, the parasite 'tolerates' her and lives alongside her. The parasite lays her own eggs, but these are looked after and cared for by the host workers as if they were their own. The host queen remains in the nest as if nothing untoward is happening and continues to lay eggs, so the host worker caste is constantly replenished. Meanwhile the parasite does not need to waste resources on workers, so only lays queen and male eggs.

In Britain, *Myrmica karavajevi* parasitises nests of *M. scabrinodis* and *M. sabuleti*; there is no worker caste. Meanwhile *Tetramorium atratulum* is a permanent parasite in nests of *T. caespitum*, but is reputed only to invade orphaned, queenless, interregnum nests. Again, there is no worker caste, which means that when the last host workers have exhausted themselves, the colony will eventually decline and vanish.

Temporary and permanent social parasitism may have evolved from the widespread ant behaviour of polygyny – having double or even multiple queens living together in the same nest. Polygyny goes against the well-known absolute monarchy of the single mother-queen ruler in honeybee and wasp nests, and which was thought (until recently) to be the norm in ant nests too. The importance, and high frequency, of ant polygyny is only just now being realised.

Under certain conditions, a newly mated queen is accepted into the already well-established nest of the same species, sometimes the birth nest from which she originated, sometimes not. Two or more queens can coexist peaceably, their offspring workers intermingling in the nest. In some *Formica* species there seem to be two different mating behaviours: some sexuals fly away into the nuptial mating swarm, while others mate on top of the nest. One theory for an evolution of nest parasitism is that this polygynous cooperation diverged such that adventurous flying dispersers flew off to make new nests on their own, while more hesitant stop-at-home queens mated with similarly circumspect males then re-entered their home nest to become secondary lodgers, tolerated by their mothers even when they started to lay eggs of their own. Eventually the behaviours were so genetically entrenched that there was no interbreeding and one form of the ant always took up lodgings in an existing colony, eventually this race becoming morphologically and physiologically distinct. Queen polymorphism is widely known among exotic ants; large winged queens fly off to mate and then attempt to make nests in new habitat patches, while small or wingless queens return to the birth nest and may eventually create their own colony by budding.

In the British Isles, budding seems to be the norm for *Formica* nest foundation. Large colonies with multiple queens have correspondingly large foraging zones, and satellite nests a short distance from the main dome are common. These can then become outlying principalities, or even independent sovereign states.

The divergence of nesting behaviours leading to speciation may be ongoing in the present day. In the western United States *Temnothorax rugulatus* shows an increasing proportion of multi-queen nests with altitude, with different rates in different geographic zones (Heinze & Rueppell 2014), perhaps suggesting that lower temperatures make returning to the home nest a more attractive option the further up the cold mountainside the ants live. Several *Myrmica* species appear to have two genetically distinct queen forms: larger macrogynes and smaller microgynes, often living together in the same nests. In *M. ruginodis* these are reckoned to simply be size variants of the same species, but microgynes found in nests of *M. sabuleti* have been designated a separate socially parasitic species – *M. hirsuta*, in which workers are either very rare or non-existent.

Unlike xenobiotic guest ants, true social parasites are usually very closely related to their hosts. This is important at least in that parasites need to be very similar to their hosts in their biochemical nest-mate recognition smell, and in any hormonal or tactile communication systems, or they would not be able to slip in and take over the colony so easily. That they might have started out as divergent behaviourally competing forms of the same species seems highly intuitive.

Dulosis

Dulosis is Greek (δουλοσις, *doulosis*) for 'slavery', and this very well sums up the ant behaviour. It's another form of permanent social parasitism, and is sometimes also called ant piracy. Perhaps this is not just because in the modern world the idea of slavery is so unpalatable that it also becomes unmentionable, but also because this particular anthropomorphism can be seen as belittling a horrible historical crime. In many Victorian books there is a definite sympathy with the notion of slavery, or at least colonial subjugation, especially if black ants are involved, and the descriptions can sound jarring and uncomfortable to modern readers. Dulosis is a natural behaviour, more akin to domestication of wild animals, but its understanding becomes clouded if we call it slavery, just as predation becomes tinged if we call it murder. It's rather a cop-out, I know, but in keeping with other myrmecological treatises I've continued to sometimes use the word 'slave', in what I claim are the interests of clarity and readability. But I do not do this lightly, and I hope I have chosen my words with care, so as not to offend or confuse.

In dulosis, an interloper queen invades a host colony, kills the resident queen and usurps the host workers to raise her own brood. These develop into males, queens and workers, but the parasite workers do not take over any work, being apparently incapable of any hard labour, and cannot even feed themselves. Instead, they adopt a piratical mode of life. Despite nautical pirates' rather romantic reputation for getting easy money from their assaults on peaceful merchant ships, dulosis is a precarious career. The pirates are a fairly useless bunch when it comes to domestic chores, and if the slave ants are experimentally removed from a test colony it quickly deteriorates.

'The slave-maker (*F. sanguinea*) attacking the nest of *F. fusca* on Shirley Common.' Frontispiece from *Ants and Their Ways* (White 1895). The common is no more, having been subsumed by Croydon's urban encroachment. *Formica fusca* still occurs on Addington Hills, seen in the background, but *F. sanguinea* is no longer there.

The pirate ants try to take over brood care, but they are pretty inept nurses. Only if the enslaved ants are returned will the nest revert to normal functioning.

Dulotic ants are, however, very effective fighters; they normally lack the short broad toothed triangular mandibles used for ant-handling food or nesting materials, and instead have long curved sharp scimitar-shaped jaws which they use to prang and dismember their victims. They mount aggressive raids on neighbouring nests of their chosen host species, kill the adult ants and bring back the host brood of mature larvae and pupae to the home nest. In this way their stock of captive worker ants is constantly replenished. In Britain *Formica sanguinea* is known to be partly dulotic, raiding nests of *F. fusca* or *F. lemani*. It sometimes resorts to temporary parasitism without dulosis, eventually forming self-contained nests which function normally, with its own active foraging and patrolling workers doing the work. However, workers of the European *Polyergus rufescens* are incapable of work, and cannot survive without their captive domestic servants. This species is widespread in France and Belgium, and since its *Formica fusca* and *F. cunicularia* hosts are common and widespread in southern England it is a good contender for colonisation here in the near future. The parasites appear to imprint with the chemical and tactile signature of their chosen host species, and once they have made a choice they stick with it, raiding new nests only of that species and ignoring other potentially suitable victims in the area. This is echoed by other species around the world. Scout ants, which explore the local terrain, recruit their pirate sisters to attack new nests, and the selection process, reinforcing one particular target species over another, is probably determined by these front-line searchers.

Strongylognathus testaceus, a very rare southern species in England, has the curved sharp scimitar jaws typical of dulotic ants elsewhere in the world. It is an obligate parasite in nests of *Tetramorium caespitum*, and it appears to raid neighbouring nests to carry off pupae in huge numbers, leading to hordes of host workers numbering 20,000 or more and vastly outnumbering the slave-makers.

Dulosis can only work if the putative pirate parasites occur in an area rich with their putative hosts. Indeed, dulotic ants never occur throughout the entire range of their host species, but only in patchy areas where their victims occur in particularly high densities. Buschinger (2009), in an extensive review of ant social parasitism, suggested that between 3 and 10 parasitised colonies per 100 host

Polyergus rufescens raiding a nest of *Formica fusca*. The long curved jaws of the raider can be seen gripping the pupa it is making off with.

colonies was a reasonable estimate (only 1.3% for inquilines). Often the host ants fight back, and parasite queens can be repulsed or killed by large, active and vigilant colonies of their hosts. There have been reports that enslaved ants might also become 'rebellious', absconding from the nest or fighting with their parasites. These revolts are never generalised through the host colony, but appear to be cases of individual ants in which some aggression mode is triggered; similar triggers also occur within conventional colonies where the occasional error in kin recognition occurs. Observations are equivocal and unclear.

In most dulotic ants, many of the captured larvae and pupae are eaten, and this predatory hunting and retrieving behaviour may be the evolutionary origin of dulosis. It is perhaps easy to imagine that in some distant antediluvian ant species a few 'lucky' potential prey pupae brought back to the aggressor's nest somehow escaped being eaten, and once they had emerged as workers set instinctively to foraging and helping in the nest, oblivious of the fact that they were now in a pirate stronghold rather than their own home. It's not quite as easy as this, though – dulosis appears not to have evolved in several ant groups which actually specialise in being predators of other ants; even though captured ant brood are sometimes stored for several days or weeks back in the nest, they are all eventually consumed. None survive.

Ant altruism and the dominance of neuters

One of the most difficult entomological concepts to grasp is the very presence of non-reproductive worker ants in the first place. Workers build and repair the nest, they find food to feed the brood, each other and the queen, they repulse competitors across the forage grounds and predators that try to invade the nest. This is familiar enough behaviour to have entered into the public consciousness, and it's frequently alluded to whenever ants are mentioned. Indeed, so well known are these aspects of ant life that they are often depicted in books and magazine stories, cartoons, films and art, even on collectable cards issued free with cigarettes or packets of margarine. But the biology behind the workers' existence is highly complex and controversial. They work themselves to death or die in suicidal defensive pacts. And yet they are sterile. They do not mate, or lay eggs – they do not make any contribution to the genetics of the colony, so how can this behaviour have possibly evolved? More importantly, how can it be passed on to future generations? This is an ongoing debate in biology, but this is my take on the subject. If it really gets too technical please don't be put off the rest of the book. Worker ants exist, even if the reasons for them are unclear.

Worker ant behaviour is termed altruism in some older textbooks; this word was originally coined (as *altruisme*) by French philosopher August Comte (1798–1857) to mean 'disinterested and selfless concern for the wellbeing of other people', as the opposite of egoism – selfishness. Comte was thinking more about the socialist ideals that build a fruitful, happy and egalitarian human society, but biologists also use this concept when considering the worker caste of bees, wasps and ants, and their apparently altruistic care of future sexual males and females, at their own personal expense.

An understanding of worker insignificance, and yet also worker contribution, has to begin with the bizarre sex-determination mechanism found in the Hymenoptera – a system called haplodiploidy. Rather than gender being determined by the X and Y chromosomes of mammals (or Z and W in birds), maleness and femaleness are founded on chromosome number. In most living organisms, the genetic material of the DNA is gathered onto chromosomes, which occur in matching pairs in every cell in the organism's body. At certain points of the growth/cell-division cycle, they can be stained with chemical reagents and seen down the microscope. This paired/doubled state is called the diploid condition. Humans have a diploid state of 46 chromosomes

Signs of the times

The existence of the sterile female worker caste of ants (and indeed of bees and wasps) has led to the odd notion of three, rather than two, genders in the nest. This vexed the ancients, who were unsure what they were. Aristotle was rather muddled about stinging worker honeybees and wasps; he thought they were males since in most other animals it was the males that were armed with weapons. But brood care was a feminine trait. It was all very confusing. The queen was variously called a 'leader', but also often a king, ruler, chieftain, prince or master and was generally considered male. And who knew what the drones were? It wasn't until 1609 that Charles Butler proposed that the honeybee ruler (and thus also the leader of a wasp or ant colony) was indeed a female queen, and this was only confirmed by dissection of the ovaries by Jan Swammerdam in 1668. Shorthand symbols for male and female, taken from alchemy and

A *Formica rufa* worker. She will nurse, forage, fight (note the dented gaster), build and support the colony. But in the end she is expendable.

astrology, had been adopted by Linnaeus for plant/flower sexes, and this soon spread to use in animal descriptions. One consequence of the Hymenoptera worker caste was the need for another symbol to accommodate this additional neuter gender. Others have also been coined.

Symbol	Meaning	Alchemical origin	Astrological significance
♂	Male	Iron (from which masculine weapons and armour are made), possibly representing a shield and spear	Mars, god of manly war of course
♀	Female	Copper (softer than iron, and major constituent of bronze), perhaps a bronze mirror with handle	Venus, god of love, what else?
☿	Worker or neuter (also used for intersex or hermaphrodite in other organisms)	Quicksilver/mercury (perhaps signifying fluidity/rapid movement)	Mercury, messenger of the gods, perhaps always rushing about. Also Hermaphoroditus was a child of Hermes (Mercury) and Aphrodite (Venus)
♃	Major worker	Tin (minor but significant constituent of bronze)	Jupiter, largest planet in the solar system
♀	Minor worker, although used in place of ♀ in some older books; for example Curtis – see page 36	Alternative version of copper	Alternative for Venus

On the whole these shorthand symbols are only used on museum specimen data labels, or in scientific tables, figures and diagrams, where they are used to save space.

'Ants at play', a touching but perhaps tongue-in-cheek notion from *The Insect Book* by Howard (1904). In fact worker ants work; that's all they do.

(23 pairs) – usually denoted as $n = 23$ (or $2n = 46$). Elephants have 28 pairs ($2n = 56$), Hedgehogs *Erinaceus europaeus* 44 pairs ($2n = 88$), and the Field Horsetail *Equisetum arvense* 108 pairs ($2n = 216$). The only time this changes is during the creation of ovum and sperm (ovule and pollen in plants), when the chromosome pairs get separated and only one of each pair ends up in each sex cell. Human ova and sperm contain only $n = 23$ single chromosomes. This is the haploid condition. However, at the point of fertilisation in the womb, ovum ($n = 23$) plus sperm ($n = 23$) combine and the full human diploid complement of $2n = 46$ chromosomes is reinstated. However, this does not happen in bees, wasps and ants.

In ants (as in all the Hymenoptera), a mated queen stores her mate's (or mates') sperm alive in a storage facility in her abdomen (the spermatheca). She can do this for years. A multiple-mated queen leafcutter *Atta sexdens* may store 300 million sperm to use over her lifetime; a much smaller *Solenopsis* fire ant stores 7 million to use over her approximately seven-year life. These, at least, have been counted; the common black pavement ant *Lasius niger* is likely to be in the same league, since queens are reported to survive nearly three decades under experimental conditions. As the queen lays each egg, she can decide whether or not to fertilise it with some of that stored sperm. Thus, she can lay two different types of egg. Fertilised eggs, with n chromosomes from the maternal ovum and n chromosomes from the sperm, become diploid; they contain $2n$ chromosomes, and will develop into female ants. Unfertilised eggs contain only n chromosomes – those original n from the maternal ovum, and none from any sperm; they remain haploid and will develop into males. Ant species across the world have varying numbers of chromosome pairs, as shown in the table opposite.

Ant chromosome numbers – what exactly is *n*?

Across their species range, ants exhibit an astonishing variety of chromosome numbers, even within species (see *Myrmica rubra* below). In an initial survey of 520 ant species a value of $2n = 20$ ($n = 10$) seemed the commonest result, with $2n = 18$ and 22 ($n = 9$ and 11) very close behind, suggesting that one of these might have been the ancestral default number at some early point in ant evolution (Cardoso *et al.* 2018). The vast majority were distributed relatively evenly across a very broad range of $2n = 16$ to 54 ($n = 8$ to 27). Here is a brief selection.

Species	Haploid males (*n*)	Diploid females (2*n*)	Geographical region
Myrmecia croslandi	1	2	Australia
Ponera scabra	4	8	Japan
Atta colombica	11	22	Central America
Lasius fuliginosus	14	28	British Isles and Europe
Lasius niger	15	30	British Isles and Europe
Myrmica rubra	23	46	Japan
Myrmica rubra	24	48	British Isles and Europe
Formica rufa	26	52	British Isles and Europe
Formica rufibarbis	27	54	British Isles and Europe
Pseudomyrmex gracilis	–	70	South America
Dinoponera lucida	60	120	Central and South America

Although the mechanisms that underlie chromosome-number changes are poorly understood, their occurrence is thought to play an important role in speciation through reproductive isolation, since ill-matched chromosome numbers (different numbers from each parent) are unlikely to give rise to viable hybrid organisms. By way of comparison, humans have a diploid $2n = 46$, reduced from the $2n = 48$ of all the other great apes, when two small chromosomes merged into one at some distant time-point way back in hominid evolutionary history. Quite how this all works in the labyrinthine twists of haplodiploidy perhaps still needs to be examined, but interest is such that an online database of ant chromosome numbers has now been established for all to see – www.ants.ufop.br. As of February 2021 there are 1,136 entries from 138 genera in 11 subfamilies.

The path of a male egg is fairly straightforward. It starts off as a haploid male egg, becomes a haploid male larva, and will eventually pass through haploid male pupal stage to become a haploid adult male ant. But what of a diploid female egg? Depending on the various poorly understood nutritional and hormonal inputs discussed earlier, the female egg will become either a sexually active queen female ant, or a 'neuter' worker female ant. Various developmental pathways for this decision split are considered later. However, haplodiploidy now offers some clues as to how a non-reproductive female worker caste can exist.

Despite the fact that workers do not pass on their own genetic material to future generations, it is, in fact, all about passing their genes on to the next generation. Even though they don't lay any eggs, workers' genes *are* passed on to future generations because they are so very closely related (they're all sisters in the nest) to the 'proper' reproductive females – the queens – that they will help rear. The idea of evolution favouring survival of the fittest is not just about individual physically fit organisms surviving and breeding again (this is called 'personal fitness', a sort of strict parental lineage measure of success). True evolutionary success is about the passage of shared genes down through the colony, or community, or species. This genetic 'inclusive fitness' is what evolutionary biologists study and measure. Even so, it still seems counterintuitive that a gene (or package of genes) producing a sterile worker caste or altruistic behaviour could have evolved, because the sterile altruistic workers still die without offspring through which they can pass the genes forward. Haplodiploidy seems to offer some answers.

Ant workers (diploid and therefore female) get half their DNA from their father, but because of the haploid nature of the male ant, they get his entire genome, all n of his chromosomes, unaltered, not subject to the usual halving or mixing of chromosome pairs seen in most other sexually reproducing organisms. Thus they get all of his genes and half of their mother's. This means that, rather than full sisters sharing an average of only 50% of their DNA with each other, it is 75%. This is also more than the 50% genetic relatedness a conventional non-haplodiploid organism (you or me, say) would have with any sons or daughters. Feed these figures into a mathematical model of ant breeding, and sure enough it is more advantageous (in terms of the progression of their genes into future generations) for the workers to help the mother-queen produce more queens and males (containing enough of those same genes), than to branch out and start a new family on their own. Biologist and mathematician J. B. S. Haldane (1892–1964) summed this up very neatly. When asked if he would lay down his life for his brother he's supposed to have replied, 'two brothers, or eight cousins'. Even though the workers do not have any offspring themselves, the workers' genes have an evolutionary inclusive fitness through their shared occurrence in the sister queens, and the sister queens ensure the lineage continues. This evolutionary success, where individuals do not necessarily have their own offspring, but their genes are still passed on through their kin, was termed 'kin selection' by theoretical biologist John Maynard Smith (1920–2004).

It has been the received wisdom for decades that haplodiploidy, with its special 75% sisterly relatedness quotient, gave rise to ant altruism. It's what I learned at the University of Sussex more than 40 years ago, where John Maynard Smith was one of my lecturers. And although it's a good place to start in trying to understand how altruism could have evolved, it now turns out that things are far from this simple. It now appears that the magic relatedness value of 75% is a bit less well defined than originally thought.

For a start, not all colonies have just a single reproductive queen in them. About half of all known ant species have colonies with multiple queens. These polygynous nests (as opposed to monogynous, one-queen nests) may contain two, three, four or more queens – thousands in large colonies of the pharaoh ant *Monomorium pharaonis*. In nests of *Hypoponera punctatissima* a single winged queen is often accompanied by numerous slightly smaller wingless worker-like queens called ergatoids. In some non-British species a few of the workers have retained the ability to mate; termed gamergates, they sometimes serve as additional or replacement egg-layers. Having multiple queens has some benefits in that polygynous colonies usually produce sexuals, new queens and males, before monogynous colonies. This is especially important in the colonisation of empty (usually unstable) habitat patches, and polygynous species are often the first colonisers of these difficult sites. But in all these cases, even if these females are closely related, they will have mated with different males, so the genetic relatedness value between the cohorts of their various daughter workers will no longer be anything near that special 75%.

Any assumed ant monogamy can also be thrown out of the window. Mating is costly and dangerous anyway for females, exposing them to predation risk out in the open, and increasing their risks of getting sexually transmitted diseases (usually fungi). Multiple mating serves only to increase these risks, and add in the real possibility of physical damage as multiple males harass her. Nevertheless, multiple mating occurs widely. In leafcutter, driver and harvester ants, with potentially huge colonies, mating with several males may help ensure that the queen never runs out of sperm for the several million eggs she may end up laying during her long lifetime. Unlike termites, where a female queen and a male king live together in the nest, queen ants are unlikely to be able to mate again later in life. It may also be that the genetic diversity gained from several fathers can increase colony efficiency and give better resistance against pathogens, parasites and environmental perturbations. But it is still genetic diversity – again diluting that special relatedness away from 75%.

Finally, it appears that supposedly sterile workers are not always actually sterile. Leaving aside the few species that can generate reproductive gamergate workers, it turns out that workers can often lay eggs, even though they do not mate. Though their abdomens do not contain a spermatheca sperm-storage organ, some still have functional if small ovaries, and they can sometimes lay unfertilised eggs which will develop into normal haploid males. Obviously this throws the personal fitness/inclusive fitness distinction into disarray, and workers 'ought' to be striving to produce their own sons to better get their genes off down into perpetuity. But laying their own eggs creates a conflict between worker sisters, because although each may want to beget true sons, who wants to rear some other worker's brats, which are only part-related nephews? Indeed, where workers do lay their own eggs, these are often removed and destroyed by others in the nest, and the offending uppity egg-laying workers are attacked. This process, beautifully described as 'worker policing', prevents the pushy few pushing their own selfish interests ahead of the colony. Nevertheless, some worker eggs do get through the police net. Meanwhile the South American 'dinosaur' ant *Dinoponera quadriceps* has no true queen and *all* the workers are capable of laying eggs, although this is usually taken on by one dominant high-ranking worker at a time. By now it is clear that the supposedly all-important 75% relatedness value does not seem to exist in the real world.

Just to nail the lid of the 75% relatedness coffin firmly shut, other oddities include: parthenogenetic ants, where queens do not need to mate, and lay unfertilised but diploid eggs that all become females; unfertilised eggs developing into queens while fertilised eggs become workers; males developing from fertilised eggs after expulsion of the maternal nucleus, so that fully fertilised eggs produce diploid males. In one of the most peculiar and perplexing (but also enlightening) discoveries, different genetic lineages of *Pogonomyrmex* harvester ants and *Solenopsis* fire ants show a genetic basis for caste creation. Even if nutritional or hormonal inputs guide an individual ant larva's development, there is now plenty of evidence (e.g. Anderson *et al.* 2008) that genes actually play a key role in determining

Collectable cards issued with Wagner's *Eigelb* (egg-gold) table margarine, *c.* 1930. Though attractive, they show varying degrees of understanding of colony life by the artist as the ants: **1** manoeuvre larvae, twigs and the dead, **2** cajole a stag beetle, **4** defend their fortress thistle flower against a flying tiger beetle, **5** stitch leaves together, and **6** carry leaf fragments but cannot see where they are going. And what the heck is happening in **3**? The caption calls it the ant fire brigade, and although the text on the reverse talks about maintaining forage paths clear of obstacles, this concept is as laughable as it is comical.

worker/queen development, though this is often in the form of a moveable predisposition, rather than a fixed unalterable predestiny. The early decades of the 21st century have certainly been exciting times to be a myrmecologist, and a review by Heinze (2008) strives to summarise the seeming chaos, at least by listing the available scenarios even if no overarching 'standard' exists.

All those mathematical models, based on haplodiploid 75% full-sister gene-sharing, start to unravel at the edges or crumble into incomprehensible dust. What is the point of it all then? The theoretical biologists will not give up without a fight, though, even if the mathematics start to become monumentally complex. There are still various ideas about how and why an altruistic worker caste could have evolved. It still hinges on the idea of inclusive fitness (success of genes rather than success of individual organisms), and kin selection (helping close relatives) is still the most powerful force in all the computer modelling. Reasonably, haplodiploidy must be an important contributor because this eusocial behaviour, with 'neuter' worker females, has evolved at least 11 times in the Hymenoptera. Conversely, eusociality has evolved only once in mammals (Naked Mole-rats *Heterocephalus glaber*), once in cockroaches (termites), once in beetles (*Platypus*), once in aphids, and once in thrips (both in communal gall-dwelling species).

A recent suggestion is that altruism evolved as an insurance policy against the volatility of the natural world (Kennedy *et al.* 2018). Cooperation, it seems, is more common (in birds and mammals, as well as in ants, wasps and bees) in unpredictable and harsh environments. Helping your relatives rear a successful generation serves not only to get your shared genes passed on into the future, but also creates a long-term genetic stability that preserves a certain status quo in the gene flow through dangerous and unpredictable times. The altruists are hedging their bets (that's another one of those metaphors borrowed from the world of economics).

What is important is not that the *number* of offspring should just be higher than others, but that there should be a better *consistent* average. Again, this flies in the face of ideas about individual 'fitness' and of getting the highest number of offspring out there ready to continue the blood line. Struggling to rear as many offspring as possible all of the time only works when times are good, but nature is fickle, and in a boom-or-bust world, that bust, when it comes, means personal extinction and the end of your line. It's a short-term policy with

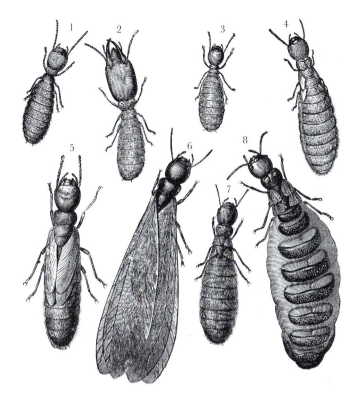

Eusocial cockroaches are called termites. Like ants they have complex caste societies: **1** worker, **2** soldier, **3** nymph, **4** ergatoid worker, **5** male nymph, **6** winged male, **7** female, **8** fully gravid female. This lovely image is from a popular Victorian encyclopaedia, *Cassell's Natural History* (Duncan 1883).

poor long-term prospects. Altruism in ants might work like an anti-entrepreneurial mechanism, actually lowering the average number of offspring, but ensuring better consistency overall. In the mathematical models, altruists that successfully rein in their recipients' reproductive volatility actually increase and spread rapidly through the population. Natural fluctuations in the nest population caused by an unstable and unpredictable environment become stabilised by sociality. This is virtually an argument for socialism. Indeed, the wealthy gentleman naturalist and myrmecologist Horace Donisthorpe would have been outraged. In the second edition of his famous monograph on ants (1927a) he bemoaned the extensive five-volume work *Le monde social des fourmis* (The social world of the ants) by his French contemporary Auguste-Henri Forel (published 1921–1923), claiming he found it 'disappointing … that the opportunity has been made for airing the author's socialistic views. I should wish in particular to protest against the ants being employed as a supposed weapon in political controversy.' He goes on (rants) at some length on this.

There can also be direct survival improvements in members of a colony willing to 'care for' each other at the level of the individual.

In the sub-Saharan 'matabele' ant *Megaponera analis*, fighting injuries are common since this species specialises in hunting termites. One in three ants in a raiding party of 200–600 individuals lacks at least one limb. The raid on the *Macrotermes* mound only takes 5–10 minutes, but the termite soldiers are armed with powerful mandibles and fight back to defend their lives, their home and their honour – which usually involves biting off the legs of their attackers. Killed termites are hurried back to the *Megaponera* nest, but the returning victorious column does not forsake its fallen. Injured ants are examined by passing workers, which respond to a help pheromone made up of dimethyl disulphide and dimethyl trisulphide. Intense antennation is followed by rescue as the damaged ant is carried back to the colony. Only ants with one or two missing legs are helped; those that are more severely damaged are left to their fate. This is not a judgement on the future usefulness (or not) of a partly dismembered nest-mate, but seems more to do with the rescuers being unable to cope with an unstable writhing patient, while they can manage with an ant that is only partially incapacitated. Back at the nest, any termites still attached to the ants are removed, and severed limbs are then licked intensely for up to an hour. This action reduces mortality to only 10%; if, however, these walking (or crawling) wounded are left to their own devices, 80% of them die (Frank *et al.* 2018). A small number will themselves become prey for other animals, as they struggle back home from the battleground, but the licking of the cut leg stumps appears to reduce bacterial and fungal infection significantly. Once healed and recovered, the five- or four-legged ants are able to adapt to a new gait and readily take up arms again on the next raid.

Another promising line of enquiry involves green beards. Please don't roll your eyes here. This is serious. Evolutionary biologist Richard Dawkins came up with this improbably fanciful notion in his famous book *The Selfish Gene* (1976) to illustrate the possibility of a gene for altruism. Imagine – 'I have a green beard and I will be altruistic to anyone else with a green beard.' If a gene coded for a genuine physical trait (such as a green beard), and that trait was recognisable by others of the species (who could fail to notice such an odd personal grooming statement?), and it engendered preferential treatment among carriers (a sort of supportive club mentality among green beardies), then it would spread through a given population. Bring in a 'spite' effect against those lacking green facial hair, and the clique of green beards soon becomes a powerful force, cooperating to support each other and down-tread conventional beard colours (who continue to live boring, selfish, non-cooperative lives). In slightly hypothetical mathematical models of social insect colonies, green beardiness can eventually dominate and become the successful norm. This remained an interesting, if speculative, idea until the publication of a paper entitled 'Selfish genes: a green beard in the red fire ant' (Keller & Ross 1998). I so wish I'd been able to have a paper with a cool title like that.

This research, on the invasive *Solenopsis invicta*, showed that some newly emerged and mated queens land from their nuptial flight, but do not necessarily have to start new nests on their own. Instead, as considered earlier when discussing nest parasites, they could be accepted, like lodgers, into previously established nests. This would lead to polygyny, with several relatively unrelated queens sharing the egg-laying, but only if they had the right genes at a particular locus, defined in the chromosome literature as Gp-9. This appears to be a gene (or series of genes) that codes for an odorant-binding protein – so not for green beardiness exactly, but part of the all-important olfactory system that ants use in kin recognition, and which they use to discriminate 'friends' which they greet with friendly altruistic behaviour. There are two forms of the gene, denominated as recessive b and dominant B respectively. Bb queens are graciously accepted into the fold, to join the nest clique and contribute to the ever-increasing socialist civilisation. Queens with B on both of that particular pair of chromosomes (BB) are spitefully attacked and killed by the existing colony's Bb workers by selective assassination. They can only survive if they move elsewhere to make a new nest for themselves and start from scratch. Queens with bb have some sort of larval problem early on in life and rarely develop through into adulthood. It only takes about 15% of the colony workers to be Bb to cause this switch. This genetic basis for altruism might also explain the existence of intermingling supercolonies.

Incidentally, the idea of a gene for altruistic or at least friendly behaviour came from two seminal papers in the *Journal of Theoretical Biology* by Hamilton (1964), when the notion of genes affecting social interaction first gained widespread interest. I like to think that science-fiction author John Brunner had come across these (he quotes another article from the same journal) when he wrote his ground-breaking and prize-winning *Stand on Zanzibar* (1968) about a slightly dystopian future world (set in 2010) where the human population of the planet is nearing critical mass and gene screening, genetic manipulation and enforced sterilisation are the norm. One strand of the complex novel is the discovery of a gene for non-aggression explaining why the then fictional nation of Benin was such a nice friendly place to live when the rest of the overpopulated world was at each other's throats. The human/ant metaphors of aggression, cooperation, exploitation, civilisation, selfishness and altruism are everywhere available to us.

Tramps, invasions and the ant megalopolis

Ants are small, and in the modern world they are easily transported around the globe hidden in commercial goods, especially horticultural material. A number of nuisance 'tramp' ant species have been accidentally spread far beyond their original geographic ranges. It's important to have a look at these invasive aliens because they offer key insights into many aspects of ant biology and ecology. In the British Isles the pharaoh ant *Monomorium pharaonis* (and its congeners *M. floricola* and *M. salomonis*), *Nylanderia vividula* and *Hypoponera ergatandria* have become established in heated buildings, some more often than others. In warmer subtropical zones these and other alien ant species can easily become naturalised, and very abundant. Rapid invasion by a new species into a new territory is well known, and is usually thought to be a result of unfettered expansion because it is taking over a new unoccupied (or perhaps ill-defended) ecological niche, but has none of its homeland predators, parasites or diseases to keep its numbers in check. Similar well-known non-ant invaders in the British Isles include the North American Grey Squirrel *Sciurus carolinensis* in the 1950s, the Ring-necked Parakeet *Psittacula krameri* from the Old World tropics in the 1960s and 1970s, and the east Asian Harlequin Ladybird *Harmonia axyridis* in the 2000s.

One of the best-known invasive tramps is the red fire ant *Solenopsis invicta*, which was introduced from South America into the United States, in the port city of Mobile, Alabama, some time between 1933 and 1945. It soon spread uncontrollably, at nearly 200km/year, and is now frequent across all the southern states from Virginia and Missouri to Florida, Texas and southern California. It is blamed for interfering with local ecosystems, ruining cultivated land, and stinging people, pets and farm animals. It has not got to Europe yet, but if it did, it would become a considerable nuisance and exert a major ecological influence on native ants and other organisms.

A series of *Les merveilles de Monde* (wonders of the world) collectable stickers issued free with Nestlé's baby milk powder, *c.* 1950. They show slightly stylised and stereotypical ant behaviours: **1** brood care, **2** foraging, **3** trails, **4** communication, **5** retrieving a caterpillar, and **6** tending aphids.

It is sometimes unclear where an invasive tramp originated. Despite being named for the land of the pharaohs, *Monomorium pharaonis* is unlikely to have started out in Egypt. Linnaeus named it in the middle of the 18th century from specimens he thought were from north-east Africa. It was probably already common in the area at the time. However, it only arrived in the nearby Mediterranean islands of Cyprus, Sardinia and Sicily around 2004 (Wetterer 2010), an unlikely late date if it really were a truly north African species. It had already been recorded from much of Europe, North America, south-east Asia and Australia more than 100 years earlier. Other cosmopolitan food pest insects had been moved back and forth across the ancient world several millennia before, along the busy trade routes of the Greek, Phoenician and Roman merchants. For some idea of its home turf, we need to find it in more natural conditions, away from the modern human buildings it occupies today. It turns out that the pharaoh ant occurs out of doors, under logs and stones, far from human habitation only in Malaysia, Indonesia and the Philippines, suggesting this is its genuine original native range, and that it made its way into the eastern Mediterranean (and elsewhere in the world) by trade with the East Indies during the days of sail.

Despite the apparent abundance and ubiquity of ants in the natural world, there are still checks and balances in their numbers. Colonies vie with each other, and around their borders they maintain a fragile peace or occasionally go to all-out war. Populations expand and contract over time, but on the whole a balance of species make-up and population numbers remains relatively constant within a given geographic area. But if a new ant species arrives from outside this well-balanced zone, it can give rise to a new phenomenon – the supercolony.

When the Argentine ant *Linepithema humile* was accidentally introduced into Europe from South America at the end of the 19th century it started to spread rapidly along the Mediterranean seaboard, and it now occurs from Portugal to Greece (Wetterer *et al.* 2009). However, the Argentine ant's behaviour also seemed to change when it got to Europe. In its native South America, neighbouring colonies of the ant compete with each other for resources and the usual skirmishes occur, keeping each nest separated from its neighbours and preventing each from gaining overall control of an area. In Europe a change in colony structure was noted. Instead of each nest operating on its own and maintaining a discrete territory against its neighbours, the nests seemed to take on whatever workers were coming through as if they were merely satellite station outposts rather than individual colonies. Neighbouring nests did not fight with each other, rather they cooperated. They exchanged workers, brood and food (by seemingly indiscriminate trophallaxis with all and sundry), and each nest (or group of interconnected nests) contained multiple queens. The colonies were no longer acting as independent city states, but as a single coordinated ant nation – a supercolony, a sort of ant suburban sprawl.

The reason for this seems to be that the Argentine ants in Europe had arisen from only a very few initial colonists. There is just not enough genetic variation in their subsequent descendants to distinguish one colony from the next. The cocktails of all-important kin-recognition chemicals by which a worker might normally distinguish friend from foe were as near as damn it identical across all ants from all the nests, right across the invading front. Inbreeding had created one big happy ant party.

Statistics now start to reach implausible tabloid headline proportions. One single interrelated Mediterranean colony appears to stretch over 6,000km^2, and contains millions of semi-individual nests and many billions of sister ants. Since the 1970s *Linepithema humile* has been a sporadic indoor species in the British Isles, apparently only able to survive in heated buildings. However, a large interrelated series of nests was recently found out of doors in gardens in west London (Fox & Wang 2016), so there is every possibility of an expanding supercolony becoming established here some time soon.

Similar invasive supercolonies are recorded for *Formica yessensis* in Japan (45,000 nest heaps, 300 million workers), *Lepisiota canescens* in Madagascar, South Africa and Australia, and *Anoplolepis gracilipes* in south-east Asia. Care has to be taken, though. Not all invasive ants form supercolonies, and not all supercolonies are created by invasive alien ants. In fact many supercolonies seem less than super. Native species can sometimes form loosely cooperating confederacies if their close genetic relatedness means they are so similar to each other in those important kin-recognition chemicals that they avoid conflict. Several species of *Formica* wood ant are known to create new nest heaps by budding the colony, a group of queen(s) and workers setting up a new nest heap nearby. Less dramatic than their Argentinian cousins, they have attracted much less tabloid press coverage and their societies are usually not described in terms of being supercolonies, but by the more staid scientific term polydomy.

Human interactions with ants

O n the whole, the British have traditionally had a begrudging respect for ants. Despite being a regular nuisance in panties and pantries, ants do not engender the same burn-it-with-fire hatred as is dished out to, say, wasps, spiders or scorpions. We have, perhaps, the curiosity of the ancients to thank for this. Though there are no true seed-collecting ants in the British Isles, it was the large and obvious *Messor* and *Pogonomyrmex* seed-harvester species that inhabited the Mediterranean region during the Bronze and Iron Ages that caught the eyes of the philosophers of the day, and which made it into an oral then a written tradition – which to this day influences much of Western cultural thought on these insects.

Though probably paraphrased, King Solomon (*c.* 970–931 BC) celebrated these seed-collectors and urged others to follow their work ethic:

> *Go to the ant, thou sluggard; consider her ways, and be wise;*
> *Which having no guide, overseer, or ruler,*
> *Provideth her meat in the summer, and gathereth her food in the harvest.*
>
> Proverbs 6: 6–8

So great was this biblical approval of the Formicidae that it was repeated a few pages later:

> *The ants are a people not strong, yet they prepare their meat in the summer.*
>
> Proverbs 30: 25

The King James Bible uses the word 'meat' in the very early 17th-century sense of 'food' or 'provisions' of any kind. At the time the Bible writings were first formalised, 1,500–2,500 years ago, hard cooperative work in the grain fields, and prudence in preparing for

OPPOSITE PAGE:
Train carriage and ants installation by scrap-metal artist Joe Rush at London's Vinegar Yard food-stall market near London Bridge.

Illustration from the quaint 19th-century natural history picture book *The Insect* (Michelet 1875), showing a not quite clear understanding of how ant columns work. This might have been a simple reworking of a picture of a human army marching in orderly form through the pass, with scouts and lookouts.

the lean of winter, would have been the high guiding principles of a rustic, rural, agrarian society struggling to cope with the vagaries of weather patterns and the constant threat of biblical locust plagues. The parallels between ant and human harvest would have been easy to appreciate. It is extraordinarily lucky for the ants that they were perhaps most often seen collecting wild seeds. They are perfectly capable of carrying the relatively large seeds of wheat, barley, emmer and spelt, and maybe they could be forgiven for taking spilled grain gleaned out in the field.

In fact ants were well known around the threshing floors, and were reported by Aelian (the Roman author, Claudius Aelianus, *c.* AD 175–235) as extorting high losses from domestic stores. He also knew that the ants chewed off the seed radicle (the embryonic root), thus preventing the seeds from germinating inside the nest. Moggridge (1873) and White (1895) both comment on instructions contained in

the Talmud which seem to indicate that pauper gleaners in the fields were able to take at least some part of any ant-stored corn grains that they uncovered in harvester nests. There also seems some evidence that these ants took enough corn to make seeking out their nests a worthwhile activity, and Landsell (1908) implies that a tithe from such uncovered stores needed to be paid. The ants did not, however, appear to climb the crop stalks in the fields to attack the ears of corn. Consequently, harvesters (Aelian describes them as 'these excellent creatures') were not considered major agricultural pests.

It was harvester ants that had also inspired Aesop (*c.* 620–560 BC) to write about them in his fables. Fable number 373 in the Perry index (named for the Illinois classics professor who researched them in the mid-20th century) is *The Cicada and the Ant*, although the non-ant protagonist is often rendered as a grasshopper, sometimes beetle or other

Mediterranean harvester ants *Messor* species coming (and going) on the trail.

The famous ants-and-grasshopper fable, here retold on a Gallaher cigarette card from 1931, using a bush-cricket and some ants of uncertain leg number.

THE ANTS AND THE GRASSHOPPER.

One fine winter day a colony of ants were busy drying grain they had collected in summer, when a grasshopper, perishing with hunger, came along and begged for food. "Why did you not treasure up food in the summer?" one of the ants asked him. "I had not leisure," he replied; "I was singing and dancing." "Then," said the ant, "those who sing and dance all summer must starve in winter."

THE MORAL. *In good times remember that bad times may come.*

Nº 18
ISSUED BY

insect. The gist is that the ants spend all summer diligently collecting food for their stores. Come the winter the cicada/grasshopper is hungry, claiming he had no time for work because he was too busy singing. The heartless ants send him on his way, chiding him that if he sang all summer, he should now dance. The latest iteration of this ancient tale is the Disney/Pixar film *A Bug's Life* (1998), in which the hard-working ants collect the harvest, but have to make an extorted tithe protection payment to the gang of lazy, greedy, aggressive, swaggering grasshoppers, which later arrive to terrorise the colony. In the end the ants rise up heroically (along with a ragtag of circus bugs, echoing *The Seven Samurai/Magnificent Seven*) to drive off the hoppers so that in future they may peacefully live off of the rightful fruits of their own land. It's all very morally uplifting and hugely entertaining – my favourite insect film without any doubt. The fact that the lead locust (Hopper: 'It's a bug-eat-bug world out there') gets captured and devoured by a bird is fine; although the ants are anthropomorphised to within a millimetre of their purple four-limbed forms, the grasshoppers are deliberately portrayed more creepy-insect-like with drab mottled camouflage colours, the correct six spiny legs, rattling wing-cases, toothed jaws, and wrinkled, horny carapace. It is no surprise that interpretations of the fable constantly intermix grasshopper and cicada, a muddle that continues to this day. The ants, though, are obviously always ants.

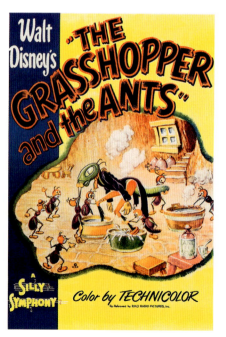

The Grasshopper and the Ants (Disney Studios, 1934). An ant colony collects an unlikely harvest of carrots, cherries and sweet-corn kernels, while a jolly, but slightly ill-mannered, expectorating grasshopper casually plays his violin. Later the grasshopper collapses in the snow outside the ants' front door, but the kindly occupants take him in, as a reformed and grateful in-house musician. In truth, ants and grasshoppers seldom interact.

The harvester work ethic had been known since ancient times. Greek and Roman writers had variously named ants *myrmex* and *formica* (nothing to do with Formica worktops, an early form of plastic, named in 1912 as an alternative 'for mica' – a mineral widely used in electrical insulation at the time). There was sometimes a bit of confusion with termites, but throughout the classical literature these small, communal, cooperating, foraging, nest-building insects were usually easily identified as 'ant' and were treated as a more or less uniform entity (Beavis 1988). Apart from the myth of giant gold-digging ants in India (the size of a fox or dog, according to Herodotus), all ants were thought to show the same harvest-type foraging behaviours, digging tunnels, storing food, organising armies, engaging in massed flying displays.

Seed-collecting scientifically observed

For many hundreds of years these tales of ant diligence and fortitude were more or less parroted verbatim, with little or no new observation, but eventually this well-worn tale started to fray. Thomas Mouffet (also Muffet or Moffet) marks about the tail end of the regurgitated granary tale in his *Theater of Insects* (1658), the first book in English on insects (though originally written in the 1580s/90s and previously published in Latin in 1634, some 30 years after his death). This landmark volume is full of proud and correct information on bees, wasps, scorpions, bed-bugs and others, but his treatment of ants is woefully ignorant. He gushes great screeds of lyrical pomposity about ant armies going valiantly into battle led by their generals, a few lazy non-fighting ants not being fed, then being tried before a council and executed 'that their young ones may take example', a hatred of grasshoppers (because they sing) and dormice (because they sleep), ants hurting elephants and bears and driving dragons mad, ants feeding on the venom from serpents and toads, but not 'fruit polluted with menstruall bloud', them only stinging idle people, misinterpreted leg-shovelling soil excavation, and some rather fanciful ideas about ant architecture involving parlours and reception rooms. Any serious mention of storing up grain (or indeed any valid description of real ant behaviours) is rather lost in the maelstrom of misinformation on much else. What is clear, though, is that as far as Mouffet is concerned all ants are the same, and the idea that different types might have different behaviours is alien to him.

At least by the time John Ray's *Historia insectorum* was published in 1710, five ant species were recognised in England; these he describes (in Latin) as: two small (one black, the other red), two medium (one shining black, the other red) and one large – the only one he names, *hippomyrmex*, 'the horse-ant', implying powerful size just as horse chestnuts and horse mushrooms are larger than non-horse varieties. These same five are more explicitly named (and can more or less be allocated to modern species) in the first book in English wholly dedicated to the Formicidae, *An Account of English Ants*, by William Gould (1747): the small black ant [*Lasius niger*], common yellow ant [*Lasius flavus*], jet ant [*Lasius fuliginosus*], red ant [*Myrmica rubra*] and hill ant [*Formica rufa*].

At this time in the 18th century there was sometimes outright dispute that ant granaries even existed; maybe Aristotle and the rest of them had seen ants carrying building materials for their nests? Or perhaps the 'harvested grains' were really the pale ant cocoons that sometimes get transported about if a nest is inundated, or under attack? By the end of that century this would become the received wisdom of the age. Shaw (1806) states dogmatically: 'This care of the ants in conveying their pupae from place to place seems to have been often mistaken for a sedulous industry in collecting grains of wheat, which the pupae, on cursory view, much resemble.'

Gould's book (1747) marks a halfway stage in the discussion. He quotes all those previous accounts from Solomon, Pliny, Virgil, Horace and Aldrovandi, and even a recent report from the French Academy, stating that ants do indeed 'have magazines, do lay up corn, and despoil it of the bud, in order to prevent it growing'. Gould quite clearly knew different ants, and as well as the five British species he was prepared to accept the presence of many other different types around the world. He thought that no British ant collected seeds, but suggested that differences in climate might explain why Mediterranean ants stored grain or seed food for their still-active colonies during the milder winters, but northern European ants needed nothing because they shut down into deeper winter torpor. This was actually a very good ecological assessment. His little book makes fascinating reading, and it seems obvious that he had made pretty close personal studies of ants. He accurately describes their body structure (although he quaintly uses the term 'saw' for their serrated mandibles rather than 'jaw'), identifying the standard 12-articled antennae, triad of ocelli, one- or two-segmented pedicel, and the easy shedding of the queen's wings compared to the male's. He knew that some (*Lasius* and *Formica*) larvae spun ellipsoid silk cocoons but those of *Myrmica* were naked, that ants moved their brood up and down the nest mound according to the weather, and that foragers came back to the nest with their abdomens distended with juices, though the aphid honeydew link had not yet been made. His oddest conjecture is that some ants nest in molehills. It is possible that he had mistaken the loose earthen tumbles made by *Lasius* species for the spoil heaps of the almost certainly ant-eating moles, but it turns out that *Formica exsecta* nest-budding can, indeed, take advantage of the diggings of moles (Bliss *et al.* 2006). Having said this, John Clare (see page 231) wrote often about molehills as seats and cushions for country folk, though he was quite clearly describing *Lasius flavus* ant hills (Heyes 1993), so perhaps this muddle was a common misconception at the time.

Despite Gould's measured analysis, the fundamental stored-grain debate rattled on for a century more. Ignoring rumours from the unreliable south, the then ant intelligentsia of the time was firmly established in northern Europe, and none seemed able to fully contemplate the veracity of those ancient reports. There is something about the tone of the Reverend William Kirby's prose (he being the more religious of Kirby & Spence 1818) that suggests he needed to have the biblical authority of Solomon and the ancient gravitas of Aesop and Aelian confirmed, or at least assuaged, in a north-west Europe which was obviously free of such harvesting ants. He seems resigned to the fact that no British ants store grain, but could not quite bring himself to suggest the Bible was wrong. Perhaps Solomon was merely emphasising the obvious cooperative foraging that was going on? Clearly more work was needed.

It was not until 1873 that J. Traherne Moggridge finally confirmed, to the presumed relief of all those northern myrmecologists desperate to hang on to the old beliefs, that the large *Messor* ants he had seen in Menton, on the French/Italian Riviera, definitely stored seeds in excavated granaries 'about as large as a gentleman's gold watch' (that's a measure I'd like to see return into common usage). He watched the ants' behaviour, strewed small ceramic trick-seeds in their path, dissected out their nests and identified the seeds and dried fruits of more than 12 different plant species, from at least seven different plant families. He also confirmed radicle-chewing behaviour, reciprocal nest-raiding, drying of wet seeds in the sunshine, and the fact that they would also dismember and devour a dead grasshopper if it was offered to them. This obscure book is a fascinating example of then

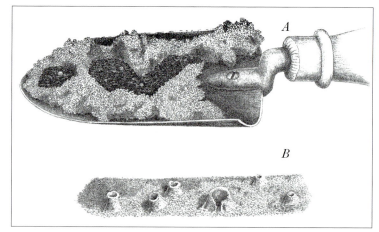

Plate from Moggridge's (1873) book *Harvesting Ants and Trap-Door Spiders*, showing his choice of nest-excavation trowel. Below are the raised entrance holes to the nest.

cutting-edge research. Moggridge carefully lists 16 species of Riviera ants and comments on several Indian ants mentioned by his various correspondents. Many of these he had personally observed carrying seeds, including *Myrmica* (now *Tetramorium*) *caespitum*, the only British species known to regularly feed seeds directly to its larvae.

Myrmecology in the scientific age

With this nugget of natural lore now confirmed, myrmecologists were set to move on to new studies. Over the next century and a half, ants moved on from being simple yet admirable examples of an agrarian society, the topic of fables and moral storytelling, and became popular research tools. Part of this popularity stemmed from the ability to collect a few ants, hopefully a queen and enough attendants to ensure her comfort, and house them in a formicarium (or formicary). Simple glass jars, wooden boxes or moated platforms could easily be set up to contain the ants and prevent them escaping, while their intimate behaviours could be closely observed. Indeed this was already a widely practised art. In 1669 Jan Swammerdam described a simple set-up:

> *I provided a large deep earthen vessel, and about six inches from the brim or verge of it, I put a bank or artificial rim of wax, and then on the outside of the circumference of this I poured water, in order to prevent the ants confined in this enclosure from getting out. I afterwards filled the cavity of this dish with earth, and therein placed my little republic of ants.*

(From the English edition, 1758)

He was able to watch his ants lay eggs, the hatchling larvae grow and pupate, and eventually emerge as adults. Different soil substrates could be offered, humidity and moisture levels could be experimentally manipulated, diverse foods could be presented. Tellingly, Swammerdam was unable to tell what the captive ant grubs were eating; he offered them sugar, raisins, apples and pears, but was never able to rear them through to adulthood unless worker ants were also present in the confines of his earthenware formicary. This was wholly different from the easy process of rearing butterfly and moth caterpillars simply by sticking in a few leaves of food-plant and leaving them to get on with it. The idea of trophallaxis would have been a wondrous revelation to him.

Later, other researchers would build more elaborate nest boxes with glass sides to observe tunnelling and brood-rearing, monitor larval growth rates, calculate colony numbers and biomass necessary for queen/drone generation, or engineer territorial disputes with other captive colonies. It was in one such laboratory formicary that German entomologist Hermann Appel reputedly kept a queen *Lasius niger* for nearly 29 years (Kutter & Stumper 1969) – seemingly the longest-lived insect on record. Ants make very good pets, and even today school or household formicaria can be bought or contrived for the education and moral advancement of our children. See page 320 for suggestions, including a simple home-made version.

With increasing technological advances the field of ant biochemistry has gradually revealed the Formicidae as masters of alchemical bio-sorcery, through an understanding of their pheromones, venoms and other physiological substances. Since John Ray first experimented with spirits of pismires, effectively distilling formic acid, and Bonnet ran his finger across an ant trail, wiping away the trail pheromone, many hundreds of ant chemicals have been isolated, identified, synthesised, and their effects quantified in the field and in the laboratory. Apart from personal allergic reactions to stings, or the displeasure of having formic acid sprayed into your face, these studies are rather outside the orbit of most British field naturalists.

Observing ants in action is a more universally accessible scientific discipline. The 19th and early 20th centuries were geared to trying to understand ant behaviours, and although some broad natural history could be noted out in the field, the professionalisation of entomology hinged on making it more of a laboratory-oriented occupation. Formicarium designs proliferated. By a strange coincidence, it was parallel advancement in medicine (and the laboratory-style observation available in medical institutions) that fuelled the nascent science of myrmecology. Auguste-Henri Forel (1848–1931) was the foremost ant specialist of his era, but his day job, from 1879, was professor of psychiatry at the University of Zurich Medical School. This was a fashionable scientific field at the time. Forel worked on melding the study of human brains and psychologies with ant brains and their behaviour. In medicine he is considered one of the founders of the neuron theory, the concept that the nervous system is made up of discrete individual nerve cells that interlink, communicating between each other to create a psyche – thoughts, character, conscience and behaviours – and that the whole is greater than the sum of the parts. Parallels with the simple, sometimes ritualistic ant behaviours made them attractive animal models to use in trying to unravel complicated human drives, wants and needs. To myrmecology Forel is the father of colony and social studies. His most famous book was the monumental five-volume *Le monde social des fourmis* (The social world of ants, 1921–1923), and his *Les fourmis de la Suisse* (The ants of Switzerland, 1874) was just one of more than 280 other scientific publications on the Formicidae. He even named his home La Fourmilère (the ant colony). Forel's contribution to entomology cannot be overstated, and although Donisthorpe (1927a) chaffed at some of Forel's more socialist views, it is no surprise that, despite this, his chapter on ant behaviour is actually titled 'Psychology'.

Another ant behaviourist trailblazer was William Morton Wheeler (1865–1937), professor of applied biology at Harvard University. Wheeler reported widely on the extremely complex and diverse behaviours of ant species from across the globe, and their close comparisons with human behaviour. This was a time when human social behaviour (external interactions between other people, as opposed to internal psychological drives), under the heading of sociology, was becoming increasingly popular. It was not just that ant societies acted a bit like human societies – showing social care and cooperation, centralised civil engineering projects and imperial expansion aspirations. This was also a time when human social behavioural constructs were being analysed and attempts made to try and break them down to link them to more basic animal origins. Just as humans had physically evolved from arboreal ape-like ancestors, so too it was proposed that human social mores such as cooperation, altruism, charity, respect, family life, religion, moral codes, love and

Auguste-Henri Forel photographed about 1900, when he was perhaps the leading ant researcher in Europe.

monogamy must have evolved from more primitive animal instincts – though please, nobody mention cannibalism, incest or promiscuity. At the time, neo-Darwinism was trying to make sense of community behaviour in terms of the advantage to a lone individual struggling to survive in the midst of it all. Sociologists argued that rather than individuals constantly competing with each other and the world at large (that original Darwinian notion of personal survival of the fittest), mutually advantageous social interactions were actually the norm, whether this was in human societies, ant colonies, flocks of birds, herds of wildebeest, or even 'solitary' animals coming together to mate and form temporary family groups.

Behavioural studies continue to dominate ant research. More recently, economic and computational problem-solving have proved very useful as metaphors and models when trying to understand how strings of relatively simple individual ant acts have evolved into landscape-altering behaviours. Analyses of ant communication systems, their mass interaction and colony organisation demonstrate the wonders of fairly basic instinctual behaviour. It is a wholly anthropomorphic idea that ant behaviour is driven by some anticipated end point such as 'build a nest', 'attack that other colony' or 'store seeds for winter'. In reality behaviour appears to be broken down into simple formulaic responses such as 'if temperature is X then move brood up-tunnel', 'if chemical Y is present then exchange food', 'if trail pheromone is present at concentration Z then follow the trail'. These simple either/or algorithms combine, just like lines of computer code, to elicit complex and apparently rational coordinated activities – and indeed, ant behaviours are often visualised using the complex flow-diagram style beloved of programmers and systems analysts. It is possible that individual ants have something in the order of 20–45 behavioural acts – elementary decisions based on local stimuli and relatively small amounts of information. Despite claims of 'learning', of ants being 'taught' to negotiate mazes and 'solve' navigational problems, there is a great difference between the way ant and human brains work. This is best summed up by the, I think, rather sad chapter heading in Hölldobler & Wilson's book *The Ants* (1990), where they claim 'Ants do not play'.

Another conceptual leap in ant study comes with the idea that an ant colony can be considered as a superorganism, with individual ants regarded more like the unit cells of an organism's body, rather than as individuals themselves; the colony can be viewed as a single animated entity, rather than a collection of independent animals. This is a very useful analogy in ants, since the vast majority of those in the colony – all those neuter workers – are outside of the usual evolutionary processes of competition, survival and passing on genes to their offspring, since workers don't have offspring. Much of ant behaviour, from foraging and nest expansion to warring and nuptial flights, can be analysed and understood using the superorganism as a model. An excellent exploration of this concept is given in another book by Hölldobler & Wilson called (not surprisingly) *The Superorganism* (2009), although the concept was first formulated by Wheeler in 1911. That ants work together as one is obvious to see, even in casual wildlife encounters in the garden as black pavement ants disperse on their mating flight, or many small workers gather together to ant-handle a large prey item back to the nest. This is not to suggest that there is any metaphysical overarching colony-wide control going on in the collective ants' brains, but that, just as in Forel's neuronal understanding of human brain activity, the ant colony acts with the same singular coordination as if it were a unified organism – it possesses, if you will excuse the jumbled metaphor, a hive mind.

Ant study is now mainstream in international biology, and is the complete antithesis of the rather comical idea of entomologists as frivolous butterfly-collectors. Ant items now regularly make the national news – not just persistent invasions of pharaoh ants in run-down blocks of flats, but scientific research articles on ant foragers navigating backwards with heavy prey, ant land-speed records in the Sahara, urine-drinking ants in arid Australia, the changing phenology of nuptial flights in response to climate change, and cannibal ants discovered in a Polish nuclear bunker. In 1990, Hölldobler and Wilson's *The Ants*, a book the size and weight of a paving slab, won a Pulitzer Prize in the general non-fiction category. Myrmecology is now officially a cool science.

British field myrmecology

British naturalists, particularly those specialising in bees, wasps or ants, may be mildly interested in ant physiology and ant behaviour, but this will be a real focus for only a few people. Most field entomology is still about finding and recording species, mapping their local or national geographic distributions, monitoring regional

Horace Donisthorpe (centre), sporting wing collar and insect net, photographed in 1928 with various Crown Estate staff under the 'Watch Oak'; from his book on the beetles of Windsor Forest (1939).

abundance and assessing their conservation status. To this extent, British myrmecology is still rather dominated by the towering figure of Horace St John Kelly Donisthorpe (1870–1951). Described endearingly, but also slightly mockingly, as 'eccentric', Donisthorpe was a hangover from Victorian times, and photos show him usually dressed in a dark three-piece suit, wing collar, tie, fob watch chain and pocket handkerchief. Possessed of a private income, he was very much the gentleman scholar, and after a brief attempt at medical studies at the University of Heidelberg (which he either found too arduous or too bloody, if you read into the reports of his 'too sensitive nature') he devoted the rest of his life to a personal study of ants and beetles. As a man of leisure he became an unofficial staff member at London's Natural History Museum and made full use of the reference collections and library there. He also set up elaborate breeding cages to study the relationships between ants and the many beetle species that live in ant nests. These same cages were later used by Chapman and Frohawk when they worked out the ant-based life cycle of the Large Blue butterfly (see page 252).

Donisthorpe was, without doubt, a superb field entomologist, and his discoveries in British ants and beetles are manifold. His two major contributions to myrmecology – *British Ants: Their Life Histories and Classification* (1915, 1927a) and *The Guests of British Ants* (1927b) – still stand as monumental (and useful) works. But he was mildly derided, even during his lifetime, for describing many dubious species of insect (especially beetles) new to science.

British and Irish ants – onward and upward study

The official list of British and Irish ant species changes regularly. Several factors intertwine to make any checklist obsolete almost from the moment it is published. One is the true biogeography effect of fluctuating ant faunas through Europe spreading across to the British Isles, becoming established, or becoming extinct here. This is compounded by whether any given author considers the Channel Islands as being a real part of the British Isles, or whether obviously exotic species intercepted at ports or accidentally introduced into heated buildings should be considered 'British'. Another is the constant reinterpretation, by humans, of exactly what constitutes a real species anyway, and whether one population differs from another enough to warrant giving it a separate specific status – or not. The following table describes the vicissitudes of three centuries of ant species accounting in the British Isles.

Author	Date	Number of species	Notes
Ray	1710	5	Listed and briefly described (in Latin) but not named
Gould	1747	5	English names only, but identifiable to modern species (see page 216)
Stephens	1829	20	The first genuine British catalogue
Smith	1851	18	Catalogue of specimens in the British Museum
Curtis	1854	16	Myrmicinae only, implying about 29 British species in total
Smith	1865	32	First tabular display of names, nest situations, flight periods, localities
White	1895	34	Based on Smith (1865), good full synopsis of the fauna. Also lists non-ants like *Methoca* (now Tiphiidae), *Mutilla* and *Myrmosa* (now Mutillidae)
Saunders	1896	20	At this time all putative *Myrmica* species were considered to be five subspecies of *M. rubra*. Also includes five species of introduced exotics
Donisthorpe	1915	33	Plus numerous introduced non-native species
Step	1924	27	No exotics, introduced or Channel Island species
Donisthorpe	1927a	36	Plus large number of introduced or intercepted exotics
Kloet & Hincks	1945	34	Plus 2 introduced species. Landmark checklist
Morley	1953b	27	Rather simplified list
Bolton & Collingwood	1975	44	Includes Channel Island species and 10 extra exotics
Brian	1977	46	Includes 5 known only from the Channel Islands
Fitton *et al.*	1978	42	Plus 9 introduced established aliens. National checklist
Barrett	1979	47	Provisional distribution maps, including Channel Island species
Collingwood	1979	46	Fennoscandian monograph, but clearly lists and discusses those known from the British Isles
Gauld & Bolton	1988	42	This number quoted for 'endemic species'
Skinner & Allen	1996	46	Includes Channel Islands
Pontin	2005	42	Surrey list; '30 out of the 42 native species'
Barnard	2011	≈60	Includes Channel Island species and aliens
Jones	2021 (this book)	66	Includes Channel Islands, and slightly zealous inclusion of recently split species, plus 18 exotic species of dubious status in British Isles or which might colonise soon

Almost all of these have since been erased from the lists as being merely very slightly larger, smaller or more strongly marked specimens, or in some other way different from existing species in only the most minor and insignificant of characters. He also, completely unselfconsciously, promoted the rather vain idea of renaming the well-established genus *Lasius* as *Donisthorpea* (as first suggested by Morice & Durrant 1915), something that very few other naturalists sought to endorse. Nevertheless his legacy will be that his books and scientific articles are jam-packed with information, observation and opinion, and they are still accessible and readable today. Who else would gleefully report that Mr C. Best Gardner kept a specimen of *Myrmica ruginodis* alive for 21 or 22 days without its head, that *Myrmecina graminicola* smells faintly of raspberries, that *Myrmica scabrinodis* is a 'skilful thief', or that *Lasius brunneus* is a 'cowardly, stupid and uninteresting species in captivity'?

British ants are now relatively well studied. The Bees, Wasps and Ants Recording Society (BWARS) is an active hub for news, information and records about all species of the Aculeata. It has built on the availability of a variety of useful and accessible ant identification guides since Saunders (1896) – notably the Royal Entomological Society handbook by Bolton & Collingwood (1975) and a more recent handbook by Skinner & Allen (1996). Its website (along with occasional provisional atlas publications) gives the latest updates on nomenclature and geographic distributions, and its newsletter publishes significant observations and interesting

Ant plates (he also included *Mutilla, Myrmosa* and *Methoca*) from Saunders' important work on the British Aculeata (1896).

reports. There have also been numerous influential national and international monographs. Most of these (for example Brian 1977, Collingwood 1979, Pontin 2005) were seen as forward steps, advancing the body of British ant knowledge, but one (Morley 1953b) has gone down in biological folklore as having received the most blistering book review in the history of British entomology (see *The dangers of writing books about ants*, page 240).

Ants are now on a roll, and as well as the book lying humbly before you, recent European guides by Seifert (2007), Czechowski *et al.* (2012), Blatrix *et al.* (2013) and Lebas *et al.* (2019) have brought field study of the Formicidae to the entomological fore such that ant appreciation is now available to all.

Ants in history, literature and art

Despite biblical authority, ants are first mentioned in literature not as earnest harvesters diligently collecting grain to the shame of onlooking sluggards, but as a tribe of bloodthirsty warriors. According to ancient Greek myth, Hera, queen of the gods, sent a plague to kill the population of Aegina, because it was named after one of Zeus's lovers. This seems a very tenuous reason to enact genocide, and the island's King Aeacus was understandably upset. He prayed to Zeus (coincidentally also his father) for some way to repopulate his kingdom, and since the local ants were unaffected by Hera's disease, Zeus transformed them into humans – hence they became the Myrmidons (or Myrmidones/Murmidones). This slightly supernatural race were as fierce and hardy as ants and wore brown (leather?) armour in homage to their origins. They later followed Achilles in the Trojan War, as detailed in Homer's *Iliad*. Their heroic status was slightly decayed in pre-industrial Europe when 'myrmidon' came to mean something akin to modern 'robot' – a hired underling or ruffian who executes the commands of his master without question or scruple.

Ants are equally represented in many other cultures – with varying degrees of sympathy. In Chinese lore they are often celebrated as suitable examples of righteousness, orderliness, patriotism and subordination (Matthews & Matthews 2005). In the Hindu tradition they represent the transience of existence, although in Zoroastrianism they represent dark forces and are seen as the enemy of agriculture. In Aztec Central America, Quetzalcoatl transformed himself into an ant to steal maize, but this is represented as part of the discovery/ origin myth of the region's staple cereal, rather than as a nuisance pest invading the pantry.

Ants don't seem to appear in ancient Egyptian hieroglyphics (though there are bees or wasps), but they do turn up in many medieval illuminated manuscripts. Apart from the now overly familiar biblical/fable/Disney cartoon notion of hard-working agrarians carrying grains, the ant's sense of smell was supposedly linked to the Christian's ability to distinguish orthodoxy from heresy (Sleigh 2003), but ideas of them working together (like men, who should be working in unity with each other and with God) are also frequently alluded to. Some of the earliest depictions appear in the Theological Miscellany (*c.* 1236–1250, MS3244, British Library Harley Collection), the Northumberland Bestiary (*c.* 1250–1260, MS100, the J. Paul Getty Museum) and the Peterborough Psalter and Bestiary (*c.* 1300, MS 053, Cambridge, Corpus Christi College Library). The illustrations often show a series of blob-shaped blobs, with or without variable numbers of legs, and they could be ants, termites, woodlice or almost any other small creepy-crawly.

Realistic representations of ants make only a slow appearance during the Renaissance and into more modern times – perhaps because they are less aesthetically pleasing than butterflies, dragonflies, beetles and other large, attractive and showy insects. Jan van Kessel the Elder (1626–1679) is renowned for his quirky insect paintings, usually on copper panels used as decorations on the cabinets of curiosity of his wealthy patrons. Many of his butterfly, moth, dragonfly, hornet and bumblebee depictions are easily identifiable to species; there are a few ants and some of the larger

Ants (probably) in the *Northumberland Bestiary,* an important illuminated manuscript from about 1250–1260, now in the J. Paul Getty Museum in Los Angeles. It is one of the more ant-like representations in this genre, and with heads of wheat to suggest harvesting.

are guessable, but smaller specimens are vague and unformed. A few years later the German-born naturalist explorer Maria Sibylla Merian (1647–1717) came to the fore through her famous studies of gaudy plants and insects from her travels in Dutch Surinam in 1699. She is credited with being among the first European naturalists to observe (and sketch) insects directly, and her published pictures show a vibrancy and lifelike quality previously unknown. Her army ants are slightly more stylised than her butterflies, beetles and spiders, but show them attacking bush-cricket nymph and spiders, latching together to form their bivouac chains, and also with winged males and females, minor and major morphs.

The still-life and flower paintings of 18th- and 19th-century Europe show many insects, perhaps to demonstrate the skill of the artist, but partly to tease the eye of the beholder. *A Vase of Flowers* by Margareta Haverman, 1716, depicts a brash array of blooms visited by bumblebee, Red Admiral *Vanessa atalanta* and hedge snail, but closer examination shows several slim ant-like insects exploring the petals and stems. And although they are small and hidden, the ants decorating similar pictures of flowers by the prolific Dutch painter Rachel Ruysch (1664–1750) are just as anatomically correct as some of the other brighter bauble insects like chafers and dragonflies.

BELOW: Painted panel by Jan van Kessel the Elder, dated 1653. Although Large Tortoiseshell *Nymphalis polychloros*, Elephant Hawk-moth *Deilephila elpenor*, mason wasp, longhorn beetle and grasshopper are clearly identifiable, his ants are rather sketchy and not completely anatomically correct.

At a similar time, Japanese and Chinese artists valued small invertebrates as useful motifs, and identifiable ants regularly appear in paintings and woodblock prints, and on decorative objects such as ceramics, netsuke and sword art. These are often accompanied by a playfulness – the ants being eyed up by a hungry bird (*100 Chrysanthemums with a Bird and Ants*, Bairie Kono, *c.* 1890), as a jarring juxtaposition to ripe fruit (*Persimmon and Ants*, Shibata Zeshin, *c.* 1880), or textural versatility (*Katydid and Ants*, Mori Shunkei, *c.* 1820).

Sometimes ants are selected to give an uneasy creepy feeling, like those in Salvador

Dali's eponymous *The Ants* (1929), or more secretly hidden among his melted pocket watches in *The Persistence of Memory* (1931). Aged about five, Dali found a wounded bat and put it in a pail overnight; in the morning it was covered in ants and he put it in his mouth, ants and all, and bit it in half. Orwell (1944) obviously thought Dali was deranged, but suggested it was these bat-devouring ants that Dali is remembering as one of his regular necrophilic motifs. At least Escher drew his nine red formicine ants with precise crisp draughtsman-like quality as they crawled endlessly around his *Möbius Strip II* (1963). We may never know if this iconic illustration was inspired by real ants, but clusters of blind army ants (African *Dorylus* species), which instinctively and doggedly follow the leader, can get themselves into a terminal and ultimately suicidal circular trek up to a metre in diameter. This is known as an ant mill – 'ant death spiral' would be a more apt term, as many thousands march themselves to exhaustion, and all are dead after a day and a half.

Nowadays ants serve two main figurative purposes – to illustrate the weirdness of insects, or to offer a border-like or labyrinthine string of repetitive creepy-crawlies. My Cowichan-style bug jumper, knitted for me by my partner, has a row of ants around the waist, following each other as if on the forage trail. A quick Google search for 'ant trails' reveals a host of illustrative or diagrammatic linear drawings only loosely based on real life, alongside photographs of genuine ant columns. Surprisingly, ants feature on postage stamps rather often, particularly on those issued by some of the more exotic locations where large and startling species occur. A word of advice if you are going to search eBay shopping sites for ant stamps – you

In Britain, ants have appeared at least three times on Royal Mail stamps, although only once showing a native species. These are, left to right: unidentified leafcutters in a London Zoo celebratory issue (2000); the enlarged head of *Gigantiops destructor* for the Bristol Wildscreen Millennium exhibition (2000); *Formica rufibarbis* as part of the UK species recovery programme (2008).

will have to sift through plenty of stamps showing anteaters or from various Antarctic Territories before you get to the formicid core.

The elegant verse of refined poetry rarely descends to so prosaic a critter as the ant, but John Clare (1793–1864) appeared to be writing from personal experience when he offered his lovely *The Ants* (c. 1830):

> *What wonder strikes the curious, while he views*
> *The black ant's city, by a rotten tree,*
> *Or woodland bank! In ignorance we muse:*
> *Pausing, annoyed, – we know not what we see,*
> *Such government and thought there seem to be;*
> *Some looking on, and urging some to toil,*
> *Dragging their loads of bent-stalks slavishly:*
> *And what's more wonderful, when big loads foil*
> *One ant or two to carry, quickly then*
> *A swarm flock round to help their fellow-men.*
> *Surely they speak a language whisperingly,*
> *Too fine for us to hear; and sure their ways*
> *Prove they have kings and laws, and that they be*
> *Deformed remnants of the Fairy-days.*

I'd guess he may have been referring to the shining black *Lasius fuliginosus*, which often nests in rotten trees (though wood ants are more likely to drag bent stalks), and though they have a queen rather than a king, and are in fact fellow females, not men, they will gather together to bring back large scavenged prey items that a single ant cannot manage on its own. Clare is sympathetic, rather than put off by their massed actions, when he wonders at their government, and if they communicate by some language too whisperingly fine for us to hear (they do); and he evokes a familiar sentimental nostalgia by linking them to far-off fairy myth.

Japanese Edo period poet Gyodai (1732–1793) likewise showed a sympathy to the local ants in the face of the inevitable monsoon deluge:

> *Nowhere to go;*
> *the dwellings of ants*
> *in the summer rains.*

Perhaps he had seen them rafting, or harvesters drying their seeds in the sun after the flood.

Elsewhere ants are mentioned seldom. There are a fair few poetised retellings of the Aesop fable, and in a poem of 1877 the American quaker John Greenleaf Whittier rather sanctimoniously invokes the puniness of ants as a solemn King Solomon marches out of Jerusalem but wisely and graciously steers his horse around the ant hill rather than crushing its occupants – to the wonder and awe of the onlooking Queen of Sheba. There are a few mentions in the comic rhyme of Edward Lear, Ogden Nash and Dr Seuss. And the occasional limerick (Jones & Ure-Jones 2021):

> *A wood ant was called to defend*
> *The nest heap she made with her friends.*
> *She squirted out acid*
> *Until she was flaccid,*
> *By curling her tail round the bend.*

Apart from some fairly nonsensical lyrics by Adam and the Ants, and a dark husky biology lecture on 'sanguinary' ants taking slaves by Tom Waits, ants have never made terribly popular subjects for pop music. They invaded Ella Fitzgerald's pants in 'Bewitched', and 'Ants Marching' by Dave Matthews simply echoes ants as the tiny nothing Myrmidon robots all doing the same thing in the same way. I think this is the same fugue in 'Empire Ants' by the Gorillaz, although mention of blowing, soaring and warm air might also be trying to elicit ideas of flying ant days. The same old themes really.

Ants do make fairly regular appearances in 'modern' literature, they being the first and most obvious trope whenever seething alien masses are required. After successes with an antagonistic moth ('A moth – genus novo', 1895), Martians (*War of the Worlds*, 1898), arachnids ('Valley of the spiders', 1903) and giant wasps (*Food of the Gods*, 1904a), it was almost inevitable that prolific science-fiction writer H. G. Wells would come up with 'Empire of the ants' (1904b) in which a large Brazilian ant species evolves advanced intelligence and becomes even more troublesome and aggressive. Apparently the early Portuguese settlers had already dubbed Brazil 'Kingdom of the Ants'. This was at a time when regular reports of army ants (*Eciton* and other groups in the Americas) and driver ants (*Dorylus* groups in Africa) started to permeate through to the wider public consciousness; they were sometimes called legionary ants, perhaps to make them sound more heroic. Some observations were more scientific, such as the reports from their Amazon adventures by

Alfred Russel Wallace (1853) and Henry Bates (1863). Others were slightly more sensational. On his famous quest to find the source of the Nile, Livingstone's tent was invaded and they bit him so badly that he had welts all over his body; his servants saved the day by smoking the ants away with grass fires and then picking the biting insects from his body (Blaikie 1880).

Perhaps my favourite, though, is *The Ants of Timothy Thümmel* by Arpad Ferenczy (1924). The book itself is unspeakably turgid, as it recounts the mad ramblings of its eponymous adventurer returning from Africa where he discovered leaf markings, later deciphered as hieroglyphics recounting the history of a prehistoric ant race. Beginning with the ancestral founders Mye-Mye and Nye-Nye, with Kye-Kye, Tye-Mye, Kye-Rye and a whole host of other lame-sounding ant characters thrown in, it eventually ends with an epic battle between the 'Lord-ants' and the 'Holy-ants'. By comparison, hobbits, orcs and ents seem wholly relatable. What makes the book so interesting, though, is the 60-plus pages of enthusiastic ecological, taxonomic and behavioural notes on ants supplied, deadpan, as a commentary by Evelyn Cheeseman and Horace Donisthorpe.

Ants can be entertaining, and the recent (2015) film adaptation of Marvel's Ant-Man for the big screen brings with it visual jokes about the puniness of ants, yet their astonishing load-carrying ability, combined with the benefits of using a team of Myrmidon robot-like invertebrates to carry out a heist. Occasionally ants offer a more spiritually powerful message, as in T. H. White's *The Once and Future King* (1958), where Wart, the future King Arthur, gets transformed into an ant by Merlin and, thank goodness, soon learns the obvious dangers of national (fascist) socialism.

Meanwhile stereotypical ants continue to rear their roughly triangular heads in non-specialist and popular culture. Perhaps one of the most iconic examples is in the horror/science-fiction film *Them!*, which features ants, ostensibly the common *Camponotus vicinus*, two metres long, mutated by fall-out

Ant-Man on his steed, which he punningly named Ant-thony.

A HORROR HORDE OF CRAWL-AND-CRUSH GIANTS CLAWING OUT OF THE EARTH FROM MILE-DEEP CATACOMBS!

"THEM"

"This city is under martial law until we annihilate THEM!"

Kill one and two take its place!

THE AMAZING NEW WARNER BROS. SENSATION!

"THEM!" JAMES WHITMORE · EDMUND GWENN · JOAN WELDON · JAMES ARNESS

Anatomical accuracy was very low down the list of important attributes for a 1954 sci-fi horror film poster mashing up Cold War radiation fears and squirming hordes of creepy-crawlies from the catacombs.

from the first atomic bomb tests in the New Mexico desert. Released in 1954, it was nominated for an Oscar for its special effects, and although it is slightly tainted by association with a whole raft of other rather silly giant/alien insect films, mashing up invertebrate inaccuracy and Cold War radiation fears, it is often regarded as one of the best. There is some genuine entomology, provided by the film's hero myrmecologists drafted in by the military to contain the menace. The first encounter involves the human protagonists shooting off an ant's antennae to disable it; cyanide is used to attack the subterranean nest, but there is a subsequent suspenseful realisation that two queens have escaped to found new colonies. Sadly *Empire of the Ants* (1977), a late screen version of the H. G. Wells short story, is an unintentionally funny parody by comparison.

Conflicts and exploitation

Today the popular idea of ants as respect-worthy seed-harvesters has long passed, and in the broad-brush picture of public opinion they have become stereotyped nuisances that sneak into the larder or interfere with the picnic. To some extent this is a genuine change in our relationship with the Formicidae as a result of real changes in our ant faunas. Three invasive species – the red fire ant *Solenopsis invicta* particularly in North America, the Argentine ant *Linepithema humile* especially in Europe, and the pharaoh ant *Monomorium pharaonis* almost everywhere – have made real nuisances of themselves. Removed from their various original geographic zones, they have increased and spread without constraint from the usual predators, parasites and competition among themselves. This has brought them into direct conflict with humans as they have suddenly become much more noticeable in the environment, rather than continuing their discreet background secret lives that only the close-observing natural philosopher could discern. It is likely that some time in the

The big three invasive nuisance ants to humans across the globe.
Top left: the Argentine ant *Linepithema humile*;
top right: the pharaoh ant *Monomorium pharaonis*;
left: the red fire ant *Solenopsis invicta*;

future these invasives will eventually settle down, having acquired a controlling burden of predator/parasite load. In the 16th century early Spanish settlements on the island of Hispaniola were all but abandoned after the appearance of a stinging ant in huge numbers. Pleas to saints and the instigation of ornate religious processions did little to control the heathen ants. Today this 'plague ant' is thought to be *Solenopsis geminata*, now considered 'native enough' and only a minor, non-aggressive, part of the Caribbean fauna.

More recently, those big three international invasive species have assisted a burgeoning tabloid press coverage that now denigrates ants, labels them as 'pests', and legitimises wholesale destruction of individuals and colonies. Even away from tabloid hysteria, they are regularly vilified as being 'tramp' species – foreign invaders, not native, illegal having evaded border control, and therefore unworthy of anything but eviction and destruction. Killing ants is now

Somewhere, in a parallel universe, the name Jones is associated with poisoning ants.

something to joke about – 'It's only when you look at an ant through a magnifying glass on a sunny day that you realise how often they burst into flames' (Harry Hill).

Ant eradication is now one of the bread-and-butter activities of pest control companies the world over. All manner of ant baits, sprays, dusts and deterrents are available from the hardware store or online. Admittedly an infestation of pharaoh ants deep inside the concrete superstructure of a hospital can be a significant problem, with potentially serious health consequences, that needs to be dealt with. In such cases baits are normally the only solution. Poisoned food is left for the ants at strategic points in their foraging territories. Workers eat or drink down the food, but the toxins are slow-release rather than fast-acting. The unsuspecting workers take their full crops back to the nest and through the social stomach of trophallaxis they disperse the poison through the entire colony, including the egg-laying queens and the larvae. It may take a few weeks to polish off the last few lingerers, but eventually the colony will succumb.

Meanwhile in our private houses, a vague trickle of *Lasius niger* taking advantage of some spilled golden syrup in the pull-out larder is now met with a draconian nuke-'em policy, even though sprays and powders do little more than kill a few exploring foragers. Baits are often the only answer for eradicating a nest built inside the fabric of the building, but hygiene is the easiest and best solution – clearing up any food spills, carefully wiping shelves and any walls where the ant trails show that they have laid their pheromones. There is often no need to bring down mass destruction on them – just dissuade them from coming indoors. Alternatively you can follow the advice of the first state entomologist in the United States, Asa Fitch, who in 1856 reported the use of a piece of shag bark hickory wood:

> *The ants gathered upon this billet of wood in the course of an hour or two in such numbers as literally to cover it, whereupon they were brushed and shaken off into the fire, and the stick was replaced to collect another swarm; and in this mode the house was soon entirely cleared of them.*

The wood of hickory (probably *Carya ovata* in this case) has sticky sweet sap.

Gardeners are also increasingly wont to pour boiling water over, or mow down, 'unsightly' spoil heaps cast out by *Lasius flavus* when really these little insects are doing no harm to anyone. And ants tending aphids on garden plants are seen as abetting noxious pests so are fair game in the horticultural pesticide wars. It's all depressingly familiar to an entomologist – every small insect is a malignant morsel of evil animated matter to be swatted, stepped on, sprayed or boiled alive. Integral to this persecution is the fact that we as humans do not value much of the invertebrate wildlife we come across – it may sometimes appear odd or bizarre, even fascinating or awesome, but more often it is just seen as disgusting and swattable without a qualm.

Flying ant days are now just another way that ants torment humans by flying into eyes, nose and mouth, getting tangled in hair and generally being vulgarly over-familiar. Ants have no concept of social distancing. There was a time when the preparations for flying ant release from the nest (small piles of soil as the exits are readied) were seen as a portent of good weather, typical of the warm summer days when ants traditionally fly. Conversely, ants moving their brood (eggs, larvae or pupae) was seen as a sign that rain was coming. Such country lore is now rather scorned.

Ironically, ants were once valued enough to be sold at market. Asian weaver ants, *Oecophylla* species, have long been known as useful biocontrol agents in fruit trees (Van Mele 2007). A Chinese document reputedly written in AD 304 states:

> *In the market the natives of Jiao-Zhi sell ants stored in bags of rush mats.*
> *The bags are all attached to twigs and leaves, which, with the ants inside the*
> *nests, are for sale. In the south, if the gan trees* [mandarin orange] *do not*
> *have this kind of ant the fruits will be damaged by many harmful insects and*
> *not a single fruit will be perfect.*
>
> Way & Khoo (1992)

A similar biocontrol is reported by Bingley (1813), this time from Switzerland, to destroy caterpillars:

> *This is done by hanging a pouch full of ants upon a tree, the root of which is*
> *smeared with wet clay or pitch, to prevent their escape; in consequence of this,*
> *they are soon compelled by hunger to seize upon and devour the caterpillars.*

Quite what species is being discussed is not reported, but if there is any truth in this tale, they were likely *Formica*, *Crematogaster* or *Camponotus* species.

Commercially available 'ant eggs' fish food is definitely a thing of the past. This was a cottage industry based on people raiding wild ants' nests. According to Chris Goodlad, my inside source at Supa Aquatic Supplies Ltd, the traditional origin of ant eggs was the Baltic

Although no longer supplied, 'ant eggs' fish food was a staple of many a fairground-won goldfish.

Ant 'eggs'; actually the silk cocoons covering the pupae. Here in a *Lasius niger* nest.

countries. A fire was started near the nest and smoke wafted into it to drive the ants out. The 'eggs' (actually the pale egg-shaped pupal cocoons) could then be harvested without the risk of someone being bitten. I used to have some to feed to my fish; the adults that emerged from the 'eggs' were *Formica rufa* (Jones 1984). Of the approximately 250 cocoons in the small cylindrical cardboard tube, about 60 had emerged at some time in the packing process; 28 were more or less whole, with a similar number in dismembered bits. My goldfish ate pupae, adults and bits with equal gusto. Even assuming that a large nest might have many thousands of pupae, each colony could have yielded only a few dozen boxes for the hard work of digging, sieving, sorting, packing and shipping. I'm tempted to suggest that the practice might have begun when people started to regard ants as nuisance pests to be eradicated. Various sources suggest that ant nests used to be dug up to feed pheasant chicks. Even today a frequently proposed 'organic' control method to rid the garden of *Lasius flavus* is to dig up the nest and scatter the brood for chickens to mop up. Although this is pure speculation, I can just envisage some 19th-century entrepreneur dreaming up a scheme to make a few shillings by packaging up the ant pupae and marketing them to owners of the new-fangled aquaria that had recently become so popular. Bingley (1813) claims they were frequently fed to pheasants, partridges and young Nightingales *Luscinia megarhynchos*. According to Mr Goodlad, Supa company legend had it that ant eggs were more expensive than gold weight-for-weight, such was the effort needed to harvest them.

Supa ceased supplying ant eggs in 2012 when a brewed/cooked concoction of fish flake became more commercially viable and ecologically more sustainable.

There are reports (see various YouTube videos) of people collecting *Atta* pupae from nests for personal consumption as 'Mexican caviar', but again this is on a micro-scale for a very narrow private market. And at the end of the process it is likely that the targeted nests were destroyed or seriously damaged in the process – hardly a sustainable harvest to help ensure the long-term survival of the ants. Honeypot ants are a niche ethnic foodstuff that has never made its way onto the supermarket shelves, despite some high-end chocolatiers' gimmicks (Sleigh 2003), and the tradition of catching massed hordes of flying leafcutters in order to roast them or grind them to a flour is gradually fading into folk memory (DeFoliart 1999).

Wood ants have been used to flavour a quasi-medicinal 'folk schnapps' (*myrbränvin*) in Sweden, where the formic acid gathered by stoking an active nest mound was also reputed to be good for warts and scabies, and entire ant hills have been cooked and used in ant-baths (*myrbad*) for rheumatism and back pain (Svanberg & Berggren 2019). This ethnographic study also reports the supposed technique of holding an open sandwich over a *Formica* mound, to get it flavoured with the lemony formic acid, and Bingley (1813) reports 'a young gentleman [hence not some naive country bumpkin] in a wood near Gottenburgh in Sweden' sitting down on an ant hill and eating the ants with gusto. This set bells ringing in my brain from which I have dragged a memory from a walk in the woods long ago – 1974 at a guess. Sitting on a log for lunch, I had suggested that a simple cheese sandwich could be pepped up by the addition of a few leaves of Wood-sorrel *Oxalis acetosella*. Care was needed, though, since the flavour was provided by oxalic not acetic acid, and this was more toxic if eaten too vigorously. My picnic associate, entomologist Roger Dumbrell, then recounted how his father, or grandfather, or some other distant ancestor, used to put wood ants into his sandwiches for similar piquant effect. Unfortunately Roger died in a motorcycle crash in 1998 so I cannot go back to him for confirmation or clarification. He was the son of a Sussex farmer, and although I would still not classify him as a naive country bumpkin, he had led a semi-feral life across the chalk downs and the Weald, and had a wealth of this type of enthusiastic natural lore to recount. Thinking about it now, though, there is a clear ring of truth to his tale.

The dangers of writing books about ants: a re-appreciation of Derek Wragge Morley

The year 1953 saw the publication of (Basil) Derek Wragge Morley's *Ants*, monograph number 8 in the New Naturalist Library, a prestigious and long-running series of books covering a huge variety of natural history subjects. His book has now gone down in natural history infamy.

Morley (1920–1969) was regarded as a child prodigy. He published his first scientific article on ants aged 16, and a year later read two papers at the International Congress for Entomology in Berlin, even chairing one of the sessions. He was unconventional, and during the research for this book I unearthed a copy of a paper on 'ants' ninth sense' he wrote in 1939 (aged 19) in which he advocates the existence of some sort of ant telepathy 'that is the transference of waves produced by the brain of one ant directly to the brains of the rest of the community'. Nevertheless he went on to read natural sciences at Cambridge and settled into a life of academic study at Cambridge then Edinburgh.

Morley had a busy year in 1953. *The Ant World* was published by Pelican Books and *Ants* by Collins. *The Ant World* appeared without much remark, and indeed it was favourably reviewed in *Nature* (Hawkins 1953). The new cheap paperback format championed by Penguin (which owned the non-fiction Pelican imprint) gained another perfectly readable title. Written in a well-pitched tone, it melded the slightly chatty informal with the boffin scientific speak of the day. And it's as readable now as it was over half a century ago. But maybe the author had concentrated too much on one book, to the detriment of the other.

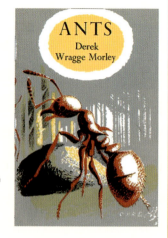

Dust-jacket illustration from Morley's book, by Clifford and Rosemary Ellis.

Ants looks, on the face of it, like the other monographs (also called 'special volumes') in the New Naturalist stable, smaller, more focused tomes, usually on a single organism (badger, mole) or narrow group (ants). An identification key with thumbnail figures and distribution maps contrasts with the anatomical diagrams, phylogenetic trees and photos of the author, pipe in mouth, fiddling with laboratory bench-top mazes or manhandling wood ant nests in the woods. But the author (though also the New Naturalist editors) appears to have lavished slightly less care on the final product, and the book is littered with silly mistakes as well as serious errors. These were brought to the fore in perhaps the most lambasting natural history book review ever to be printed. Written by fellow formicid expert I. H. H. Yarrow (1953), it contains the immortal lines:

the appearance of this volume on ants has created circumstances which defy reasonable explanation, and many readers will wonder how it came to publication in its present state. For sheer wealth and weight of errors it surely has no modern equal.

Yarrow goes on to list a litany of downright errors, grammatical faux pas, editing slips and scientific muddles ('it is like confusing a cat with a dog'); he bemoans the 'unsurpassed crudity' of the identification figures 'at times grossly inaccurate … at others completely misleading … and not infrequently ludicrous', and he mercilessly mocks the erratic and inconsistent use of 'absurd' English names.

This blistering review has sometimes led to the suggestion that the book was withdrawn and pulped, but this does not seem to have been the case. *Ants* is not particularly scarce and is easily available on the second-hand market – currently for about £10–15 lacking the dust jacket. Yarrow's review was published in *The Listener*, a magazine ostensibly about broadcast radio and television, but which also carried a broader range of book reviews, and although it was a mainstream news-stand publication, its readership would not necessarily have been a close match for the insect-book-buying public. Other reviews were mildly critical, but none quite so damning as Yarrow's. In their retrospective on the New Naturalist Library Bernhard & Loe (2015) suggest the truth is likely more prosaic. The print run of 3,900 was probably an underestimate of interest; copies shifted well, and it was sold out by the end of 1953. Yarrow's *Listener* review did not appear until 8 October 1953, by which time sales had already gone well.

Nevertheless, it appears that Collins had lost faith in Morley, and no reprint was ever considered. When the 'ordinary' New Naturalist volume *Ants* was published 24 years later, no mention whatsoever was made of the now discredited monograph (Brian 1977). Morley was obviously a bit of a maverick, and he had previously got into a spat with the

British Medical Journal over an article he wrote in *Picture Post* about a secret remedy for 'wild warts' – a manifestation of some types of skin cancer (Anon. 1950). His article was seen as thinly disguised advertising for the clinic of one David Rees Evans at the Presbyterian Hospital in Newark, New Jersey, and he was roundly denounced:

> Mr. D. W. Morley, a biologist with no medical training … Rees Evans having no scientific training himself … Mr Morley, acting as a sort of scientific compère for Mr Evans … the hopes of many sufferers will be tragically raised … Picture Post has done a grave disservice to science and to the British public … dangerous nonsense … Cancer curing is to our mind the most execrable of all forms of quackery.

Evans may well have been the quack, but Morley did himself no favours in the exchange.

Being eager and slightly naive to the point of being a bit foolish is a fault many of us have demonstrated over the years. I feel that Morley has been less well treated by history than many other credulous writers. During his short life Morley suffered serious health issues, and he died of tuberculosis at the age of only 49. Who knows whether his personal circumstances might be mitigation for a lapse of concentration in the run-up to publication? Anyway, I would urge readers to re-examine *Ants* in the softer more forgiving light of modern understanding. I recently re-read it and found it informative, if a bit quaint, even though my copy once belonged to entomologist and traveller Evelyn Cheeseman, and she appears to have made several pencil corrections in the margins to what she considered those slightly jarring errors.

Ant interactions with other species

I think we've already established that, individually, ants are small and vulnerable; they are also packed with nutrition – mostly protein-rich muscle and energy-rich fat stores. Their size does put them rather near the bottom of the food chain, and they are easily picked off by bigger and more voracious predators eager for the tasty bite-sized morsels. Their nests are also veritable caverns of cornucopia – full of succulent eggs, grubs, pupae and assorted food stores. And just as the ants benefit from the warm and dry nest interior, so too do sundry lodgers, invaders and quiet hangers-on.

Predators

Part of my garden is completely ant-free; this is down to the four free-range hens I have running around down there. Away from the chicken coop, ant enemies span several orders of size magnitude. At the top end, New World ant-bears (Giant Anteaters *Myrmecophaga tridactyla*), African Aardvarks *Orycteropus afer* and Asian pangolins are well known for digging into ant (and termite) nests after adults and brood. In the British Isles Badgers *Meles meles* wreak major havoc in the nests of ants like *Lasius flavus* and *Formica rufa*, which make large mounds full of amassed grubs. Badger excavations are targeted, well executed, almost surgical in their precision; they will also dig up wasp and bumblebee nests, and have thick shaving-brush hairs around their muzzles to protect against stings, although I can't find much information on how they cope with sprayed formic acid.

The Green Woodpecker (called yaffle where I grew up in Sussex) will eat any insect, but it specialises in ants. Less arboreal than other woodpeckers, it is a frequent visitor to garden lawns. I've seen it on mine, where it casually picks up foragers in the chicken-free zone.

OPPOSITE PAGE:
The unmistakable flat white form of the ant woodlouse *Platyarthrus hoffmannseggii*.

243

The Green Woodpecker *Picus viridis* is more of a groundpecker, after ants. Inset: Green Woodpecker dropping containing mostly wood ants *Formica rufa*.

Its droppings are highly distinctive – they look like cylinders of cigarette ash, the white on the outside being the uric acid excreta of the normal bird splash, but the centre is a stiff column of the dark undigested chitin from the ant bodies. It is sometimes possible to dissect the droppings and forensically identify which ants the bird has been attacking; the last time I checked one, Addington Hills, south London, 2016, a good supply of *Myrmica ruginodis* was obviously available. Donisthorpe (1927a) reports being sent a number of *M. rubra* taken from the stomach of a Green Woodpecker bought in Leadenhall Market in 1919 – whether the bird was purchased alive or dead we are not told; this was another era, when such things were apparently commonplace.

Everywhere, insectivorous birds, mammals, amphibians and reptiles will eat whatever small invertebrates they can find, and ants appear to make up a fair proportion of prey items whenever quantitative results are published (e.g. Razeng & Watson 2015, Mansor *et al.* 2018). Slow-worms *Anguis fragilis* and lizards are frequently found under the same stones, logs, planks and other detritus as ants, and it seems inconceivable that they do not tuck into the readily available prey.

Ants presumably make up a far higher proportion of prey taken by aerial predators during the massed flights of flying ant days.

Anecdotal observation from my back garden suggests that the Swifts *Apus apus* that scream about the neighbourhood are doing so with gluttonous abandon when the black pavement ants *Lasius niger* eject their nuptial clouds into the sky. However, a study from Poland suggests that Swifts are perhaps the least likely to feed on ants compared to House Martins *Delichon urbicum* and Barn Swallows *Hirundo rustica* flying in the same airspace (Orlowski & Karg 2013). Again, the imported red fire ant *Solenopsis invicta* provides a well-studied model in North America, where a hectare of meadowland is estimated to produce 40,000 flying queens in a single mating flight, that's 250,000 per year, and the local Purple Martins *Progne subis* were regularly recorded bringing them back to the nestlings (Helms *et al.* 2016). The authors of that study suggested that across the country the martins were probably consuming billions of the flying ant queens each year.

In North America, the crabronid wasp *Aphilanthops frigidus* specialises in capturing winged queen ants of *Formica fusca* (and several of its relatives) to stock its small tunnel nests in the ground (Wheeler 1913). It times its appearance and tunnelling activities to fit in with the nuptial flights of the ants in the area – 26 July to 16 August reported. Two or three queens are stashed per cell and an egg is laid on them. In Britain the closely related *Philanthus triangulum*

Under a piece of roofing felt laid out for reptile monitoring, a Slow-worm *Anguis fragilis* shares its shelter with a *Myrmica rubra* nest.

specialises in capturing honeybees for its burrows, but no British wasp is regularly recorded taking queen ants from the air.

Not all bird/ant interactions are predatory, though. In a still poorly understood grooming-type behaviour, birds of many different families have been observed wallowing in ant nests (Morozov 2014). This behaviour is readily observable in public parks and ornamental gardens throughout the country where the birds fluff-up and roll over in the short mown grass, the ants being easily visible crawling over the plumage. This is called, rather unimaginatively, 'anting', and there are several theories around what the birds are doing. Many birds already dust-bathe, drooping out their wings, fluffing up their feathers and fluttering/ruffling themselves in the dry dust; my chickens are very good at this. This is thought to help clean the feathers by absorbing excess oils, a bit like a dry shampoo. It may also help control ectoparasites such as lice and fleas. Anting may well be an extension of dust-bathing, and various hypotheses suggest that the formic acid may help feather maintenance or waterproofing, or it may act as a fungicide or antibiotic, or the ants could be finding and removing parasites. It may all be a prelude to devouring the ants anyway, after they have squirted out all their foul-tasting formic acid, or it could be self-stimulation – a bit like a human having a cigarette.

After birds, spiders are probably the most important predators on insects, and many take ants as a matter of course. Generalist ground- or herbage-dwelling crab spiders (family Thomiscidae) and jumping spiders (Salticidae) pounce on any small invertebrate they encounter on their forays, and equally generalist orb-web spinners such as the common Garden Spider *Araneus diadematus* take advantage of flying ant days. My father told me of a house spider, *Tegenaria/Eratigena* species, living in a small nook in the side of my parents' house – it seemed to subsist entirely on workers snatched from the forage trail of *Lasius niger* that passed close by the entrance to its lair, and their empty husks littered the ground after the resident spider had sucked them dry. There are also plenty of spiders which have developed a particular taste for ants. Many of the ant-mimicking spiders mentioned on page 62 use their ant-like appearance to get within striking distance of their ant prey. In Britain *Dipoena*, *Callilepis* and *Zodarion* species specialise in eating ants.

Being so small and vulnerable, non-flying worker ants are likely to fall victim to almost any predatory invertebrate, and there are

ABOVE: Ants make up a regular portion of the diet of ground-dwelling spiders. Here a *Xysticus* crab spider has taken a hapless *Myrmica* worker.

LEFT: A Sri Lankan assassin bug nymph, *Acanthaspis siva*, which cleverly disguises itself by gluing the remains of its victims, here at least two ant species, onto its back.

miscellaneous records of ants being taken by robber flies (family Asilidae), dung flies (family Scathopagidae), assassin bugs (family Reduviidae), digger wasps (family Crabronidae) and preying mantids. One of the most peculiar must be the Chinese spider-hunting wasp *Deuteragenia ossarium*. Wasps in this family (Pompylidae) specialise in stocking individual cells along their small burrow nests with paralysed spiders, usually ground-dwelling species, on which their eggs are laid. This *Deuteragenia* duly does, until it gets to the last cell, a small porch-like vestibule, which it instead stocks with dead ants – up to 13 specimens, selected from nine different species

Although the spider-hunting wasp *Deuteragenia ossarium* (from south-east China) stocks the cells of its tube nest with dead spiders to feed its grubs, it packs the final porch-like vestibule with dead ants, perhaps to disguise the smell of its burrow.

(Staab *et al.* 2014). This bizarre behaviour is thought to disguise the smell of its nest (which after all is full of nutritious spider bodies), discouraging parasitoids and nest-thieves, and has earned it the title ossuary or bone-house wasp.

One of the most famous and ferocious of specialist ant-predator groups are the antlions (superfamily Myrmeleontoidea). These delicate-winged insects resemble the lacewings (order Neuroptera) to which they are closely related, and the adults seem much too flimsy and fragile to be serious carnivores. But the larvae are the stuff of science-fiction nightmare. The stout, squat, slightly hairy larva has short legs, but massive curved scimitar jaws. It burrows tail-first into soft dry sand and starts flicking the grains away to excavate a conical pit a few centimetres deep, at the bottom of which it sits waiting for passing prey – mostly ants. The pitch of the sandy slopes of the pit is called in engineering parlance the 'angle of repose', the steepest angle at which a loose surface can rest without collapsing. Any disturbance, say an ant walking too close to the edge, causes the sides of the pit to start moving, inexorably bringing the ant down to the waiting jaws of the antlion, which drags its victim into the soil and sucks out its innards. If the ant does manage, by some miracle, to start scrambling up the sides, the antlion starts flicking sprays of sand up against it, using its jaws as shovels, and, undermined, the ant tumbles to its inevitable grisly fate.

The antlion *Euroleon nostras* is now established enough that it regularly turns up in East Anglian moth traps, this one in the garden of Scott Mayson. Its larva (above right) shuffles bottom-first into the loose sand where it will wait open-jawed for its ant victims.

Antlions occur throughout the world, but until recently were unknown in the British Isles apart from the odd vagrant. However, in July 1994 two adult specimens of *Euroleon nostras* were found near the toilet block of the RSPB's Minsmere reserve in Suffolk, presumably having been attracted to the lighted windows. At this point they could have been considered migrants making landfall, but larval pits were subsequently found nearby in 1996 (Plant 1998). It now seems to be well established on the East Anglian coast. Sadly, despite the enthusiasm with which this new insect is recorded and photographed, there seem to be no records of exactly which ant species it is eating. A single specimen of another species, *Myrmeleon formicarius*, was found in a moth trap on the Isle of Wight in 2013 (Cooke *et al.* 2013), so there is every possibility that this and other antlions will continue to turn up in the near future to persecute British ants. Incidentally, a good technique for observing the antlion larva is to lie down and blow very lightly into the conical pit; this gently removes the loose grains of sand, exposing the larva, which remains unaware of its nakedness for a few moments.

Ironically, probably the most important predators of ants are … other ants. These may be competing nests of the same species or aggressive predatory specialists picking off their more peaceable or under-militarised neighbours. Those desperate colonial battles described using Lanchester's power laws (page 171) are well reported in the literature as ants from one colony decimate another, taking the dead and their brood as nutritious trophies of war. As Auguste-Henri Forel succinctly put it: 'The greatest enemies of ants are other ants, just as the greatest enemies of men are other men.' Page 171 shows a specimen of *Lasius niger* I found near Bicester, Oxfordshire, on 7 September 2018; I first thought the bobble attached to its leg was a hitchhiking mite. Under the microscope later it proved to be the head of another ant, apparently hacked off in the process of biting the leg of its opponent, who was now wearing it as a gruesome badge of honour. Perhaps kneepad of honour.

Live-in predators

Simply rushing up to an ant, grabbing it, and eating it seems a bit blunt, and although predation is a very widespread nutritional model across the animal kingdom, it comes with the very real prospect of the prey fighting back as if its life depended on it. This is particularly dangerous if you are an insect not much bigger than the ant you intend to devour. Stealth now becomes the watchword. And the most stealthy sneak inside the nest and live among their prey. Two of the best-known live-in ant predators are the larvae of *Microdon* flies and Large Blue butterflies.

Molluscan *Microdon*

In 1823 German naturalist Carl von Heyden described a bizarre reticulated slug-like mollusc, which he named *Parmula cocciformis* for its supposed resemblance to *Coccus* scale insects. The following year explorer Johann Baptist von Spix reported a similar creature, *Scutelligera* (meaning 'shield-carrying') *ammerlandia*. Another, *Ceratoconcha* (meaning 'horned shell') *schulzei*, was described by Heinrich Simroth in 1907. Finally in 1924 Spanish naturalist Alejandro Torres Minguez published his account of *Buchanania reticulata*, a new 'naked snail' from the Coma de l'Orri in the Pyrenees.

All of these experienced and expert scientists had been fascinated by the strange domed, legless, slightly flanged creature that undulated very gently across the palm of the hand. Usually found under stones, or logs, sometimes in the company of ants, the classification of these lacy soft-shelled or shell-less molluscs had defied intellect. This was not really surprising – because they were not molluscs after all, but larvae of the hoverfly genus *Microdon*.

The adults of *Microdon* (about 450 species in the subfamily worldwide, four in the British Isles) are not very hoverfly-like anyway. They are drab, broad and squat, with rather short wings, they hardly visit flowers, don't appear to hover, indeed they barely seem to fly much, preferring instead to crawl or sit in the herbage around the entrances to ant nests. It is into these that the mated females will descend to lay their eggs, from which the peculiar hemispherical mollusc-like larvae will hatch.

Many hoverfly larvae are predators, attacking aphids and other small invertebrates, skewering them with their sharp beak-like mouthparts, sucking out the body contents, then discarding the empty husk of their victim. The legless maggots are often flattened, or bag-like, and they shuffle along on their bellies, testing the way ahead with the head, touching it side-to-side, just in the same way that slugs and snails feel where they are going. Despite their anatomical peculiarities, *Microdon* larvae are very similar, although they attack their chosen prey – ant eggs and larvae – through stealth. Waddling across the brood, a *Microdon* larva pierces the skin of the ant grub, empties out the body contents, then moves on to its next meal. Although near fully grown *Microdon* larvae are very sluggish (pun intended), first instars are surprisingly active and rapidly move in to attack.

A sluggish *Microdon* larva remains stationary in a frantic *Lasius niger* nest, Yealand Hall Allotment SSSI, Lancashire.

The *Microdon* adults don't get it all their own way. This one has fallen victim to the workers of *Formica exsecta*.

British *Microdon*/ant associations

Microdon species	Status	Distribution	In nests of
M. analis	Local	Lowland heaths of Berkshire, Surrey, Sussex, Hampshire, Dorset; also Caledonian pine forests of Scotland	*Lasius niger, L. platythorax*
M. devius	Rare	Mostly the North Downs of Surrey; also Hampshire and Sussex, but with outlying records from Merionethshire, East Anglia, Oxfordshire and Buckinghamshire	*L. flavus*
M. myrmicae	Local	Southern England from Surrey and Sussex to Cornwall; also Wales, Lancashire, south-west Scotland	*Myrmica ruginodis*
M. mutabilis	Very rare	Scotland (Mull and Inverness), but only recently split from *M. myrmicae*, and only identifiable from obscure larva or puparium characters	*Formica lemani*

Microdon ecology is still poorly understood. In a review of world Microdontinae, only 14 out of 43 genera were known ant associates, while the others had unknown associations, or were unconvincingly linked to termite or wasp nests (Reemer 2013). It seems that at least in some species the larva's cuticular hydrocarbons mimic those of its hosts so they can blend in unnoticed among the brood. Most *Microdon*/host associations are highly specific, with each species only visiting one ant species, or close species group (see table above).

Fiendish *Phengaris*

The Large Blue butterfly *Phengaris arion* has a fragile history in the British Isles. It was always a scarce butterfly, but at the end of the 19th century it was widely recorded from about 90 sites in six discrete regions across southern England, with its strongholds in the Cotswolds and the north coast of Cornwall. Like other blues it was associated with chalk and limestone hillsides and with luck could be found on sunny days from about 20 June to the middle of July, sitting about on the herbage (they are not very active fliers). But during the early part of the 20th century all was not well, and entomologists knew this was a declining and disappearing butterfly that needed help.

Simply fencing off a bit of land as a nature reserve area did not work. The butterflies lay their eggs on Wild Thyme *Thymus drucei* and the tiny hatchling larvae start to eat the flowers, but despite an abundance of this food-plant on many sites, the butterfly was vanishing. Closer examination was needed – on hands and knees on the hillsides.

Thyme flowers are certainly eaten, but the caterpillar is strongly cannibalistic at this stage, and if more than one egg is laid per flowerhead, only one caterpillar survives. The caterpillars continue feeding for a couple of weeks, moulting three times. But after that final skin is shed the caterpillar acts very strangely. It is still tiny (about 4mm), but it stops feeding; instead it drops to the ground, hides under a leaf or soil particle and waits to be discovered by *Myrmica* ants. If an ant approaches, the larva exudes a droplet of sweet liquid from the back of its enlarged segment 7, the Newcomer's gland. The ant drinks it down, and there follows a frenzied milking session; other ants are recruited by the initial finder, and they also lick the glandular exudations, but eventually leave the first ant still in possession. After 30 minutes to 4 hours the caterpillar suddenly rears up on its prolegs (fleshy sucker-like back legs), tucking its head down so that it resembles an S-shape. This is thought to tense the slightly flabby caterpillar's body into a stiffer consistency, closer in texture to the taught skin of an ant grub. The ant immediately picks up the larva in its jaws and carries it straight back to the nest.

In the colony the ant drops off the *Phengaris* larva in the brood chamber, carefully placing it among the ant grubs and eggs. Here it is largely ignored by the workers, although they occasionally stand over it, just as they seem to stand guard over their own offspring. The still small caterpillar spins a silk pad on which to mostly rest, but occasionally it moseys over to eat ant eggs, larvae or pre-pupae. This is not just a last-minute protein hit to ensure proper nutritional development, for 99% of the caterpillar's final body weight comes from eating the ant brood (Thomas & Emmet 1989). Despite its apparent small size, the Large Blue caterpillar is a significant predator, and during its one- or two-year occupation of the ant nest it will devour about 1,200 ant grubs. Most *Myrmica* nests are large enough to rear only one Large Blue through to adulthood, and this will usually destroy the colony. If the ants bring back another *Phengaris* larva, the combined predation load may kill the nest without either butterfly caterpillar achieving enough body weight to ensure successful transition to adulthood.

Eventually the caterpillar pupates in the upper part of the ant nest, but is still attended by ants, which lick secretions from around the spiracles on the pupa. The butterfly emerges the following June, and exits the nest attended by excited ants. As in so many insects, male Large Blues emerge a few days before the females.

A *Myrmica* ant attending a caterpillar of the Large Blue butterfly *Phengaris arion*.

Though the ant association was discovered over a century ago (Chapman 1916, Frohawk 1916), it soon became clear to naturalists that the butterfly's survival was not just linked to some limestone hillsides with thyme plants and a few ant nests. Unfortunately the precise ecological details were worked out too late to save the butterfly, and it was declared extinct in the British Isles in 1979. For an update on the butterfly and attempts to reintroduce it, see Chapter 9, page 291.

Parasites, squatters, thieves and other interlopers

Ant brood is tasty, but many non-ant invertebrates found inside ant nests are not predators. They are after other sustenance. Just as human homes are invaded by woodworm, larder pests, carpet beetles and clothes moths, so too ant nests provide specialist scavenger niches for a wide variety of insects, mostly mites and beetles. And just as human habitations are well-guarded and dangerous places for any invading insect, ant nests are well protected and relatively inhospitable. Invaders of human households have not just made a wrong turn at the kitchen door to take advantage of a few biscuit crumbs left on the plate, they are specialists at living secret lives right under our noses, and the most familiar species now occur only in our houses (Jones 2015). Likewise, ant 'guests' (Donisthorpe 1927b) usually occur only in ant nests; these are not simply opportunistic scavengers or predators, they have evolved to take particular advantage of the unique social fabric of the ant colony, and some have evolved bizarre body forms, structures and behaviours intimately associated with their lives among the ants. They are known as myrmecophiles, for their love of ants, and it has proved difficult to find out exactly how many invertebrate species around the world (or indeed just in the British Isles) are ant-nest invaders, interlopers or lodgers. Wasmann (1894) made a start by listing 1,246 worldwide myrmecophiles, Parmentier *et al.* (2014) list 125 species in nest mounds of *Formica rufa*, Hölldobler & Wilson (1990) give a 14-page table listing everything from false scorpions to froghoppers, and Parker (2016) estimated roughly 10,000 invertebrate species as being in some way myrmecophilous. Whole books have been written about myrmecophily; it is a huge subject, so what follows is not an encyclopaedic review – more a personal selection.

Turn over a log or stone almost anywhere in lowland Britain or Ireland to find a colony of *Lasius niger*, and you will almost always also uncover a number of tiny white woodlice wandering about in the ants' tunnels. *Platyarthrus hoffmannseggii* also occurs in the nests of many other ant species; in fact its host range is so broad that Donisthorpe (1927b) coined the word panmyrmecophilous just for it. This is my favourite woodlouse, and it shows several of the key features of ant inquilines – the ant equivalent of household lodgers. Firstly, it is blind, lacking any of the light-sensitive facets that even

Platyarthrus hoffmannseggii, perhaps the commonest 'guest' in the nests of British ants.

the most secretive of other woodlice normally have on their heads. Living in the constant dark of the ant colony, it has no need to see. Secondly, it is pale, lacking the usual pigments found in its relatives. Again, living in the dark underground, it has no need of camouflage, sexual signalling colours, or protection from sunburn. Thirdly, it has enlarged fatter and flatter antennae and legs compared to its non-ant-bothering cousins. Fourthly, it is broader and shallower in proportion to all other woodlouse species, with a strongly developed flange all around its body. Though it mostly lives in seeming harmony with the ants, this is probably a defence against ant attack in that, if necessary, it can clamp down on the floor of the ant corridor, and draw in its legs, presenting an unmovable bump under which the ants can get no purchase to flip it over to reveal any soft underbelly. Finally, it only ever occurs in ant nests; it is never free-living, never found wandering about on its own 'outside'. How it gets to invade new ant nests is still a bit of a mystery, but Donisthorpe (1927b) reported seeing them following ant forage trails in captive nests in the laboratory, and he also quotes Forel (1886), who observed specimens crawling along about every 25cm down the line of *Formica pratensis* workers busily moving their entire colony from one nest to another 14m away.

Platyarthrus is a true commensal (Latin *com*, sharing, and *mensa*, table, meaning to share in the same meals) in the *Lasius* nest, in that it is completely ignored by the ants which bustle past it without a second brush of the antennae. It appears to do no harm to the colony, but subsists by scavenging in the tunnels, eating bits of food dropped by the ants, including the pellets of indigestible waste material accumulated in the mouthparts of the ants (in the infrabuccal chamber) which are periodically ejected, and probably also their faecal droppings and other rubbish mouldering in corners.

There are huge numbers of other animals inhabiting ant nests. Some can be regarded as a downright nuisance, causing the same damage to an ant colony as domestic pests might in a human home (see table on page 267). Others are more or less harmless, scavenging a meagre secret life at the periphery of the ant colony. They have been variously classified according to the pet schemes of several entomologists, and although these classifications are to some extent arbitrary and flexible around the edges, the scheme originally suggested by Wasmann (1894) has stood the test of time well enough and is still useful today. It classifies ant guests into five approximate and sometimes overlapping categories.

1. Synechthrans

These are mainly predators of ants, or scavengers on the dead. They are treated with hostility by the ants and rely on speed and agility to avoid the vengeance of their hosts. Donisthorpe (1927b) called them 'hostile persecuted lodgers' (that reminds me of someone I knew in my student days), but marauding bandits might be nearer the truth. The vast majority are rove beetles (family Staphylinidae) which are highly active, mobile and manoeuvrable insects. The shortened wing-cases of the rove beetles give them great flexibility in their mostly long thin bodies, allowing them to insinuate themselves into tiny cracks, or between soil particles. They may also protect themselves with defensive secretions which the ants lick, or which become smeared on their mouthparts and antennae. *Zyras* and *Pella* rove beetles curl up their tails if challenged by host ants, secreting defensive chemicals from the tergal gland located on the upper side of the tail, between the sixth and seventh segments. These chemicals appear to show clogging, repellent or soporific qualities, and while the ant cleans itself, becomes disoriented or loses interest, the beetle makes its escape.

This is a very large group of insects, and also includes some outliers like *Drusilla canaliculata*. This pretty and highly distinctive brown and red rove beetle is very common in Britain and Ireland; it occurs under logs, stones, planks of wood and other dumped rubbish, and although not necessarily in their nests, it regularly appears near the ants also occurring under these items. It looks very ant-like,

The common and widespread *Drusilla canaliculata*, a typical ant-predator rove beetle. It's tempting to imagine that this one lost the final segment of its left antenna in a jostle with an over-defensive ant.

The flat flanged form of *Amphotis marginata* allows it to pull in legs and antennae and clamp down onto the substrate if the ants get wise to its ulterior motives.

matching them in speed and gait and curling its tail up when it dashes away at top speed. It has been found well away from ants, under rotting seaweed, in moss and decomposing haystacks, where it probably feeds on other small insects, but has been directly observed snatching injured ants and making off with them. Staphylinids are notoriously difficult to identify, but this is pleasingly one of the few which can be named in the field. When I showed one to a non-entomologist naturalist he was fascinated, and collected a specimen to take back to his office. I thought my outreach had gained a convert to beetle study; but, no, he wanted to show it to his colleague, called Drusilla. Ah well, maybe she would develop an interest in myrmecophiles.

A famous synechthran (though very rare in England) is *Amphotis marginata*, which lurks near the forage runs of *Lasius fuliginosus*. It is active at night when it patrols the trail, stopping the occasional ant and tricking it into regurgitating food by tapping it with its antennae. Hölldobler & Kwapich (2017) nicknamed it a highwayman. Well-fed ants are easily tricked, but eventually the beetle is discovered and attacked. Being broad, flat and flanged (similar to *Platyarthrus*), it simply clamps down hard to the substrate and the ants, unable to get hold of leg or antenna, eventually get bored and move off.

2. Synoeketes

Called by Donisthorpe (1927b) 'indifferently tolerated lodgers', these are mostly scavengers like *Platyarthrus*, but some are opportunistic predators. They are usually ignored by the ants because they match the host ant smell or are odour-neutral. This is another very large and diverse group of invertebrates, some of which show closer or more distant interaction with ants. When I first visited Windsor Great Park, in Berkshire, on 28 April 1985, I was very much following in the footsteps of Mr Donisthorpe, since this was his favourite hunting ground, and he made many of his beetle and ant discoveries here. I was accompanying other more experienced entomologists who knew the area well, and they were soon able to show me a specimen of the tiny (2mm) squat rove beetle *Batrisodes delaporti* in a red rotten birch log home to a colony of *Lasius brunneus*. Both ant and beetle were discovered new to the British Isles from this locality. And although the ant is now widespread in central and southern England, the *Batrisodes* is still only known from this area of Berkshire. Donisthorpe recorded it as eating the ant brood in one of his formicaria.

Perhaps because of its habit of nesting in rotten wood, many synoeketes are recorded with *L. brunneus* (and also *L. fuliginosus*), since peeling off bits of dead bark is a standard beetle-hunting technique whereas excavating and dissecting active ant mounds is a much more troublesome and rather specialist procedure. Thus, when we stopped for lunch in Windsor that day I gobbled down my sandwich quickly and examined a large oak log, idly picking off bits of the thick gnarled bark. Rove beetles dominate the ant lodger fauna, so when I found a small dull brown weevil I thought it was just sheltering there. I popped the specimen into a tube, and it was only much later, when I showed it to my colleagues, that it was pointed out to be *Dryophthorus corticalis*, again then only known from Windsor, discovered there new to Britain by Donisthorpe, and usually only found in company with *L. brunneus*. Unlike the *Batrisodes*, nobody seems to know what *Dryophthorus* eats. It's one of only a handful of weevils (superfamily Curculionoidea) known from

The enigmatic myrmecophile weevil *Dryophthorus corticalis*.

ant nests, and yet this is one of the most diverse beetle families on the planet; most of them are plant-feeders. It is possible that it feeds on the rotten wood in which the ants are coincidentally nesting, and elsewhere in the world others in the genus are simply regarded as detritivores in fungoid trees.

Many years later (29 May 2016) when I checked a large mouldering oak trunk at Lesnes Abbey Woods, in north Kent, plenty of *L. brunneus* workers rushed about, and there in the midst of them all was another tiny squat synoekete rove beetle, *Trichonyx sulcicollis*. Unusually, this species was *not* first discovered in Britain by Donisthorpe. He does, of course, comment on it, and one of the observations he makes is that it had been found by 'evening sweeping'. This arcane technique of using an insect sweep net in the calm still of the early evening, as the sun is setting, is different from the normal practice of frantically trawling through the long herbage during the day. Many beetles, especially, only fly at this time – perhaps a defence against predation by birds, which are resting up ready for roosting, and at the same time avoiding bats, which have not quite taken to the air. They are often beetle species which would not normally be flying about – wood-boring, ground-dwelling or semi-subterranean species – unlike the very many flower-visiting, leaf-feeding, dung-eating or aerial hunter species which are on the move all day long. The implication is that they have almost consciously chosen this quiet time of day to make the dangerous journey from one secretive site to another – to find new food, locate a mate or found new colonies. It illustrates that even tiny beetles which normally spend all their time deep in ant nests must sometimes venture out in the wide world, otherwise how could they find new ant nests to colonise?

And how might an indifferently treated lodger myrmecophile discover a new ants' nest? There is no experimentally verified answer, but given the olfactory dominance in ant senses, and the importance and diversity of their chemical signals, it would seem likely that their guests also use smell to find their hosts. I sometimes wonder what my rucksack smelt like that a specimen of the ant-nest scavenger/predator *Myrmetes piceus* landed on it in Bedgebury Forest, on 21 July 2016. This tiny hemispherical beetle lives with the red wood ant *Formica rufa*, and a small nest was just a few metres away.

3. Symphiles

These animals are the most closely incorporated into the nest. They are true guests, appearing to be completely adopted by their hosts as if they are part of the colony, and are not attacked or removed; in fact they are welcomed by the ants and treated like one of the family. They are fed by the ants through trophallaxis, and may offer their own secretions in return for nourishment.

Only five true symphiles are known in the British Isles (see table below). They all have a strange structural feature in common – tufts of hairs (trichomes) sprouting from glandular pores, usually on the upper surface of the abdomen. The host ants spend some time licking these trichomes, and in laboratory studies of non-British species the ants carry the beetles about by grasping them. The chemical secretions from the trichomes are highly complex, but seem to contain appeasement signals to prevent aggression, recognition signals so that nest-moving beetles can be adopted by new hosts, and finally defensive secretions – just in case.

Despite the ants' acceptance, verging on encouragement, these symphile guests are not quite neutral in the nest. Although the theft of liquid food by trophallaxis is probably minimal, they all also (as both larva and adult) attack and eat the ant brood. How the ants survive this level of predation is not fully understood, but the larvae of *Lomechusoides*, at least, are cannibalistic, so they probably self-regulate their population numbers inside the ant colony.

True British symphiles

In Britain, according to Donisthorpe (1927b) and Skinner & Allen (1996), there are only five truly symphile ant 'guests', all rove beetles (Staphylinidae).

Symphile	Status	Distribution	In nests of
Claviger testaceus	Very local	Widespread but scattered in England; also north Wales and south-east Scotland	*Lasius flavus, L. niger, L. alienus, Myrmica scabrinodis*
C. longicornis	Very rare	Isle of Wight, Surrey, Berkshire, Oxfordshire and Glamorgan	*L. niger, L. mixtus, L. umbratus*
Lomechusa (formerly *Atemeles*) *emarginata*	Very local	Widespread but scattered in England	*Formica fusca* (summer) and *Myrmica* species (winter)
L. paradoxa	Rare	Southern England, Cornwall to Surrey and Kent	*F. fusca* (summer) and *Myrmica* species (winter)
Lomechusoides strumosus	Rare	Southern England, Surrey, Middlesex, Berkshire and Gloucestershire	*F. sanguinea*

A *Lasius flavus* worker examines *Claviger testaceus*, a true ant guest beetle found only in ant nests, where it is housed, fed and protected by its ant hosts.

Claviger testaceus is widely reported as being the commonest and most widespread of this group, but it has always been a great frustration to me that I could never find it. I still haven't, but I continue to roll over logs and stones examining the galleries of ant nests, trying not to be visually overwhelmed by the swirling masses of workers and hoping that the slightly less frenetic gait of the tiny rove beetle will catch my eye. There are several recent records from the chalk scarp of the North Downs. I was planning to walk the North Downs Way, and intended to stop slightly more often than is usual on a long-distance hike, to examine a few ant hills more closely. However, the 2020 and 2021 lockdowns put paid to my plans. I'll wait. *Claviger* solicits trophallaxis food from its hosts and also feeds on the ants' eggs, larvae and pupae, or scavenges on dead adults. It is frequently groomed by the ants, which lick the trichomes at the base of its abdomen. It is rarely treated aggressively by the ants, in which case it feigns death; the ants then pick it up by a notch, the antebasal sulcus, just behind its very short wing-cases, but they soon become bored and drop it.

All of these true nest guests are rare, or rarely found. The only one I have ever discovered was a single specimen of *Lomechusa emarginata* on Addington Hills on 13 June 2012. It was under a small log housing a colony of *Formica fusca*, the summertime host of this bizarre beetle.

Away from the British Isles the pan-tropical Paussinae ground beetles are an extremely important group of symphiles. They are short, squat, flat beetles with bluntly truncated wing-cases, short flattened legs and shortened antennae often composed of a small number of broad flange- or antler-like segments covered with trichomes. About 800 species are known – all associated with ants. Many have sound-producing stridulatory organs which are suggested to make similar noises to those of their hosts (Di Giulio *et al.* 2015). The adults solicit food by trophallaxis from the ants, but also graze on the brood.

Cerapterus pilipennis, an African paussid beetle showing typical broad flat form and extreme antennae. Although none occurs in Britain or Ireland, this is an important symphile group in the tropics.

4. Ecto- and endo-parasites

These might be regarded as 'conventional' parasites, living on or inside their hosts' bodies. The severity of the 'attack' ranges from licking bodily secretions to biting a hole in the ant cuticle to drink the haemolymph or penetrating the ant and eating it alive, from the inside. Although beetles number high in terms of insect species invading ant nests, they are equalled and probably exceeded by mites. Microscopic and poorly studied, mite numbers are probably greatly underestimated. Hölldobler & Wilson (1990) provide a list of groups, but the life histories are mostly unknown. Many seem simply to be phoretic, hitchhiking on the ants, many are probably scavengers, but some at least are true parasites, latching on to an ant and sucking haemolymph out through the soft intersegmental membranes.

One slightly infamous mite is *Antennophorus grandis*, one of several species that cling tight to the body of a *Lasius* worker, either stealing food as it is exchanged between sisters, or soliciting its bearer to regurgitate for it. A single mite attaches itself to the underside of the head of a worker, but two mites cooperate to maintain the ant's balance by grasping one either side of the gaster. A third mite can then take up a head position, a fourth on top of the gaster.

Internal patasitoids of ants are even less studied than nest guests. Many flies and parasitic wasps are known to attack ants, but little is understood of their behaviour or life histories, other than a few host/parasitoid rearing records. This is an area of study still ripe for field entomologists to engage in. One of the best-known British species is the tiny (1mm) *Pseudacteon formicarium* (family Phoridae), which hovers over the *Lasius* ants on the trail before darting down to inject an egg into the gaster of one of them. Donisthorpe reports how this fly would also bob down to strike the open hand that held live ants a few

Ant under attack, in this case a North American *Camponotus* worker being buzzed by an *Apocephalus* parasitic fly. Despite a serious trawl of image libraries, no photo of a suitable British species could be found, suggesting this is still an open area for research.

moments earlier, suggesting that it can detect the formic acid or some other ant scent. It is very easily overlooked; the only time I ever saw it was when I was sitting dejectedly on a pile of road chippings waiting for a tow truck to recover my car, which had rolled off the road into a ditch on the steep chalk scarp of Saltbox Hill near Bromley – something to do with torque, traction and those loose chippings. There were no other living things to watch except the trail of *Lasius niger* ants being tormented by these flies. The ants are not oblivious of their attackers – they stop, and angle their bodies in what can only be described as a 'full alert' pose, then make a dash for it. The flies, however, are very quick, and I watched several touch-encounters where I assume an egg was laid, or at least attempted.

A closely related phorid genus, *Neodohrniphora*, similarly attacks the large workers of *Atta* leafcutters in South America, trying to lay an egg into the ant's large muscle-filled head capsule. On the way out onto the forage trail the ants are very wary of their assailants, relying on speed to outrun them, or snapping jaws and waving front legs to ward off the unwanted attentions. But on the way back the ants are laden down with large leaf portions in their jaws, and are particularly vulnerable. To try and protect themselves, the large labouring workers recruit small minims to ride shotgun on the leaf portion as it is carried back to the nest. The minims are too small to warrant attack from the flies, but offer a good defence against the parasitoids.

5. Trophobionts

This final group of nest associates includes mostly aphids and other plant bugs, and also some butterfly caterpillars, which live intimately with the ants but do not obtain nourishment from them. Instead the ants encourage and protect them in exchange for honeydew or other liquid secretions. Most trophobionts occur away from the nest (described as extranidal by Donisthorpe) and are the well-known cattle-like herds of aphids which the ants tend and protect. However, there are some that are found in the nests too (intranidal).

The root aphids tended by *Lasius flavus* (numerous species including *Geoica utricularia*, *Forda marginata* and *Tetraneura ulmi*, Ivens *et al.* 2012) are typical intranidal trophobionts, but this is a moveable feast, quite literally. Although most of the aphid species milked for honeydew are away from the nest, almost free-living on the herbage nearby, there is every tendency for ants to take things back home with them. This is borne out by the many observations of butterfly caterpillars and pupae being found in ant nests. Unlike the caterpillars of the Large Blue, which has evolved that elaborate life history with *Myrmica sabuleti*, eventually feeding on the ant brood, several other butterfly larvae are tended by ants on their food-plants, and are often taken back to the nest to pupate. These include, in the British Isles, Green Hairstreak *Callophrys rubi*, Purple Hairstreak *Favonius quercus*, Brown Hairstreak *Thecla betulae*, Silver-studded Blue *Plebejus argus*, Common Blue *Polyommatus icarus*, Chalkhill Blue *Lysandra coridon* and Adonis Blue *L. bellargus*. The precise physiological relationships are unclear, but it is likely that the larvae produce nutritional secretions that the ants find attractive. This is still an uncertain area, but there is plenty of anecdotal evidence to suggest that the larvae and pupae summon ants by squeaking – the caterpillars by muscular contraction forcing air through the trachea, and the chrysalis by rasping abdominal segments together as a row of pegs on one segment grinds against a series of grooves on the other (Thomas & Lewington 2010).

A less obvious relationship also exists between ants and several leaf beetles, notably the scarce but widespread *Clytra quadripunctata*. The adults occur in spring and summer on various trees and shrubs in the vicinity of *Formica* wood ant nests (and other ant species elsewhere in the world). Female beetles hang over or near the nest, dangling from a leaf using the four front legs to grip, while shaping a cocoon of frass (excrement) around an egg with the hind feet. The

egg is then dropped and is collected by the ants and taken back into the nest. Naked eggs, lacking the frass coat, are eaten. The *Clytra* larvae are thought to feed on plant debris inside the nest, and new adults emerge the following spring. A similar, if sometimes less closely ant-associated lifestyle also occurs in the related genus *Cryptocephalus*. These so-called pot beetles have larvae that live inside a bag-like or pot-like shell made from their own frass. They feed on dead leaves under their various food-plants. Many *Cryptocephalus* species are reported as occurring in ant nests, having been taken back there by their hosts, but others have been captive-reared in simple ant-free containers. It is possible that, like the hairstreaks, they obtain some additional rather than obligatory protection from the ants. A review of ant-associate leaf beetles is given by Agrain *et al.* (2015).

A type of satellite obligate associate category also has to be created to understand the odd relationship between ants and the 'ant ladybird' *Coccinella magnifica*. Extremely similar to the very common Seven-spot *C. septempunctata*, *C. magnifica* is rather scarce in the British Isles – the official English name is rather mundanely the Scarce Seven-spot. It only occurs near (but not in) the nests of *Formica* wood ants. It climbs on the herbs and shrubs in the vicinity of the nest, where it preys on the very aphids which the ants are attending, and which they are ostensibly protecting from such predators. *Coccinella septempunctata* is successfully attacked by the ants and quickly retreats, but *C. magnifica* is generally ignored, presumably because of some cryptic or defensive scent it gives off (Sloggett *et al.* 1998).

Ant guests, ant pests

At a stretch, the very many non-ant invertebrates living inside ant nests can be seen as analogous to the equally abundant creatures that invade human habitations. They are usually after the same things – food and shelter. They are tolerated (or not) on a similar scale administered by humans, ranging from 'kill it now!' through flicking it out of the window, to not even noticing it was there at all, or actively encouraging it. These five British ant guest species are offered as examples of the five Wasmann (1894) categories.

Guest/pest	Biology	Human home analogy
Synechthran (hostile persecuted lodger)		
Myrmedonia humeralis (a rove beetle)	Catches live ants or scavenges on the dead, relies on speed and agility to attack, then escape. Very many other insects use this strategy.	Wolf or hyena, dashing in through the window to snatch the vulnerable or the sick. Thankfully mostly in Grimm fairy tales.
Synoekete (indifferently tolerated lodger)		
Dinarda maerkeli (a rove beetle)	Larvae feed on dead ants in the midden, adults also rob ants of food by inducing them into false trophallaxis. Large number of similar scavengers do this too.	Mice scavenging in the larder, or the compost bin, or sneaking crumbs off your plate.
Symphile (true guest)		
Atemeles emarginatus (a rove beetle)	Lives among the ants, which feed it by trophallaxis. They treat it as one of their own. The most highly adapted ant guest behaviour; few similar species.	A changeling: the fairies swapped it with a real baby at birth, but the parents are oblivious to the deception.
Parasite		
Pseudacteon formicarium (a fly)	Lays eggs on adult ants; hatching grub burrows into the body and eats the ant alive from the inside. Numerous flies share this behaviour.	South American screw-worm fly lays eggs in open wounds and maggots burrow into the flesh. The stuff of nightmares, but thankfully uncommon.
Trophobiont		
Paracletus cimiciformis (an aphid)	One of many aphids that only occur in ant nests; they suck root sap and are protected by the ants, which milk them for honeydew. Also often eaten.	Dog, reared, fed and sheltered by the host family, which uses it for protection, hunting, companionship, draught, play, guidance, status symbol. Not milked; rarely eaten.

Origins of myrmecophily

Just as the invaders of human homes represent many different groups of organisms, come from different parts of the globe, and use various techniques to insinuate themselves into our lives, so too ant guests (and trespassing interlopers) have had many different evolutionary routes into the nest environment. There is, however, a recurrent theme. A recent review of ant-nest beetles by Parker (2016) makes a very good argument that certain groups, if not predestined, were at least overly facilitated in their invasion of the usually well-protected and hostile environment of the ant nest. Although 33 (about one-fifth) of beetle families contain one or more myrmecophilous species, three narrow groups have come to dominate the ant-nest fauna in terms of species diversity worldwide (also reflected in British species) – these are rove beetles (Staphylinidae) in the subfamily Aleocharinae, rove beetles in the subfamily Pselaphinae, and clown beetles (Histeridae) in the subfamily Haeteriinae. Clues emerge by considering members of these groups that do *not* live with ants.

The broader members of these three groups are all relatively small insects, used to pushing through the leaf litter and loose soil layers, and which could easily make the ecological jump to living in leaf-pile or loose-soil ant nests, where their small size would also help them avoid detection by their hosts (myrmecophilous beetles tend to be smaller than their host ants, with a mean body size of only 3mm). They are mainly predators attacking a wide range of other small invertebrates, so a change to attacking and eating ant brood is not a difficult hurdle to overcome (by contrast leaf beetles, weevils and longhorns, all hyperdiverse plant-feeding groups, are virtually unknown in ant nests). Non-myrmecophilous species from the three dominant groups specialise in finding animal dung, carrion, rotting fungus or other decaying organic matter where they target fly larvae. These microhabitats are small and very transient resources in a huge environment, which the beetles need to find quickly, usually by detecting airborne scents, so a move to detecting the smell of ant nests is a simple shift.

Aleocharine rove beetles not living with ants already have glands on the upper surface of their flexible hind bodies, between the sixth and seventh segments. These produce repellent chemicals (often quinones) that they use to defend themselves. Parker (2016) suggests that there has been an evolutionary shift in manufacturing output, from defensive repellents to ant-appeasing substances. It's as if they already had the developmental circuitry necessary to create and adapt glands and glandular secretions to this new purpose. Looking at the broad range of ant-associate aleocharines, there is already a spectrum of myrmecophily. At one end *Drusilla* nearly lives in ant nests, picks off stragglers, and uses its defensive gland secretions if it is challenged. At the other end are the true guests *Lomechusa* and *Lomechoides*, where the glands are now fully ant-attractive.

Another advantageous feature of the rove beetles' narrow flexible tail is that it has allowed many species associated with army and driver ants in both Old World and New World tropics to evolve a narrow myrmecoid waist and bulbous 'gaster', so that they actually resemble the ants among which they run (Seevers 1965). This may offer them protection against external predators, helping them to blend into the seething forage trails rather than standing out ready to be targeted, as well as acceptance by their host ants. This body form has evolved numerous times in different rove beetle lineages – at least 12 according to Maruyama & Parker (2017).

Pselaphine rove beetles, on the other hand, are all squat tough little creatures with, for their size, extremely thick integument. Those species not living in ant nests have distinctive foveae (large dimples) in the thorax and abdomen – pit-like indentations which reinforce the body, somewhat like the corrugation that makes thin metal or cardboard much more rigid. Inside the body cavity these troughs are further reinforced by internal struts and buttresses. This was perhaps an advantage against the crushing weight of the soil through which these tiny beetles crawled, but in some species it has been repurposed against the crushing mandibles of the ants with which they live. Small ant-nest pselaphines challenged by their hosts simply curl up, play dead, and wait for any futile biting attack to end. In true ant guest

Ant-mimic rove beetles that run with army ants: *Ecitocryptus*, associated with *Nomamyrmex* (left) and *Pseudomimeciton*, associated with *Labidus* (right).

species like *Claviger*, the dimples are fewer since reinforcing corrugation is no longer required, and instead they now form nooks either side of a carrying-handle grasping notch (the antebasal sulcus) which the ants use to delicately pick up and move the beetles about. This is also the area, on the first three visible segments of the hind body, which is now lined by trichome-covered glands. Under the microscope these hairs are often seen to be coated with a pale waxy substance rich in lipids, and presumably containing ant-appeasement chemicals.

Finally, histerids are built like little tanks. Again, this may originally have been an adaptation to burrowing in soil, under dung and carrion. Irrespective of size they are broad, domed, lentil-shaped beetles, with flat legs and antennae that can be pulled tight in to the body where they rest in ready-made grooves so no part of them is accessible to the biting mouthparts of any attacking ants.

These combinations of small size, flexible or reinforced body form, and pre-existing glandular structures appear to have given these three particular groups of beetles a head start when it came to inveigling themselves inside ant nests.

Ant/plant relationships

Apart from the obvious interactions of tropical leafcutters cutting leaves, ants the world over exert a sometimes more than subtle influence on plants. Indeed, many ant species have evolved quite intimate relationships with particular plants – which are referred to as myrmecophytes. The spiny acacias of Africa and South America (but not, oddly, those in Australia) are home to ants like *Pseudomyrmex*, which chew small holes to get inside the hollow bull's-horn spikes. Here they make their nests. The leaves of these acacias are tipped with small nodules called Beltian bodies, named after the naturalist explorer Thomas Belt (1832–1878) who first observed the ants' close relationship with the plants. The Beltian bodies are rich in lipids, sugars and proteins and are grazed by the ants, who take back portions of them to feed to the grubs. In return for shelter and food, the ants appear to protect the acacias from browsing animals, large and small. If the shrub is brushed by human or farm animal the ants swarm out and sting vigorously. Bushes which have had their ants experimentally removed suffer increased herbivory from sap-sucking plant bugs, leaf beetles, caterpillars and twig-boring jewel beetles. Meanwhile on neighbouring ant-infested trees all of these

OPPOSITE PAGE:
Panamanian acacia ants
Pseudomyrmex feeding
on the plant's extrafloral
nectaries. The double
'bull-horn' spines are
hollow, and the ants chew
a hole through them to
make small nests inside.

insects are attacked and removed. Similarly, ant-free acacias are prone to be overshadowed or overgrown by other plant species, but these are controlled and kept in check if the ants are present. Oddly, Panamanian acacia ants *Pseudomyrmex satanicus* did not notice the spider *Eustala oblonga* as it rested by day, often right under their eyes, hard up against the hollow ant-occupied thorns – so long as it did not move (Garcia & Styrsky 2013). Perhaps there was no need to attack it, since the spider harmed neither the ants nor their plant home – it spun an orb web at night to catch flying insects. Complacency is hardly an evolutionary advantage though; the jumping spider *Bagheera kiplingi*, which occupies acacia bushes in Mexico and Costa Rica, avoids the guarding ants by being watchful and agile, but in an apparently unique twist of fate it *does* eat the leaves, the only known plant-eating spider.

Although no native British plants have Beltian bodies, ants regularly visit extrafloral nectaries – nectar-secreting glands found on the stems and leaves of more than 90 different plant families. Received wisdom is that these plants suffer less herbivory because they are attended by aggressive ants which see off caterpillars and plant bugs. Results are often equivocal; Bracken *Pteridium aquilinum* is sometimes reckoned to be a good example of patrolling *Myrmica* and *Formica* ants being paid by the plant to protect it, but exclusion experiments suggest their impact is insignificant (Heads & Lawton 1984).

Plants regularly offer insects nutritional packages in exchange for assistance – pollen and nectar are the payment for insect pollination. This usually relies on insects flying from one flower to another to transfer the pollen, and wingless ant workers are likely only a minor part of this equation. In arid regions of the world, where ants are a major part of the ground fauna, a few plants have adopted ant pollination. These flowers usually have short and prostrate forms, with easily accessible nectaries (Hickman 1974), but they offer only meagre amounts of pollen and nectar; this is thought to discourage self-grooming, which ants seem to be particularly good at compared to other pollinators. In Spain the rootless, leafless, stemless plant parasite *Cytinus hypocistis* is pollinated by several ant species (de Vega *et al.* 2009), but specific ant pollination is not recorded in the British Isles. An Australian orchid, *Leporella fimbriata*, is pollinated by winged male bulldog ants *Myrmecia urens*, which are tricked by visual and chemical cues into trying to mate with the flower (pseudocopulation), thus ensuring pollen packages are glued onto the body for onward

A black ant *Lasius niger* feeding at extrafloral nectaries on the leaf of a cherry tree.

transport (Peakall *et al.* 1987). This seems to be a very rare behaviour. Nevertheless British ants are frequently seen visiting flowers; I would suggest ivy may get some benefit from their attentions.

The hollow thorns and stems of acacia, in which *Pseudomyrmex* ants nest, are called myrmecodomatia (usually shortened to domatia, singular domatium), and across the world similar ant-friendly spaces vary from hollow roots, leaf pouches and pithy stems that the ants excavate, to swollen branches and inflated flower stalks. Perhaps the best known is the tropical Asian epiphyte genus *Myrmecodia* (about 25 species known). The root system, anchored to a tree branch, supports a large swollen bulb, several centimetres across. This bulb is honeycombed with a labyrinth of tunnels and chambers in which *Iridiomyrmex*, *Camponotus*, *Crematogaster* or other ant species take up residence. The plant provides a sheltered nesting site for the ants, and in return the ants provide it with nutrition. Small warts inside parts of the domatium tunnels absorb nutrients from the droppings, discarded prey remains and assorted nest detritus dumped by the ant inhabitants. Elsewhere, smooth non-absorbent tunnels provide safe brood-rearing chambers for the ants.

Meanwhile, those grain-harvesters, the ones so enthusing Solomon, Aesop and J. Traherne Moggridge, do not merely inflict seed loss on the plants around them, but can alter the make-up of whole plant communities. Harvesting ants cannot carry all the seeds they discover

Sliced open swollen stem of *Myrmecodia* epiphyte, showing the ready-made chambers in which ants nest.

back to the nest, and they rarely consume all those they do successfully transport. Many are dropped en route, many are left abandoned in the ants' granary stores; if the ants forget to chew off the radicle, some seeds start to germinate in the nest and will be ejected. A seed taken by an ant is not inevitably destroyed, and it now appears that many plants actively encourage ants to take their seeds away.

Referred to technically as myrmecochory, seed dispersal by ants is a widespread strategy. Seeds from a huge array of plants, right around the world, are made attractive to ants by the provision of a relatively large nutrient-rich appendage, girdle or cap (the aril or elaiosome) on which the ants feed, usually leaving the embryo and internal food stores of the seed intact. As the ants take the seeds away, the plants can benefit in a number of ways. The seeds are more widely dispersed than if they simply dropped to the ground. Parental competition with the seedlings is minimised if the seeds have been transported elsewhere. Seeds carried into ant nests may better escape regular bush fires in the area, and are taken out of sight of other seed-feeders like birds and rodents.

Seed removal by ants is at its most significant in dry desert environments. Here, seeds may be the most frequent life form of many plant species. Accumulating in the soil and remaining viable for some time, they form seed banks ready to develop if and when the right circumstances of moisture and temperature arise. Ants have a disproportionate effect on some plant species, especially 'preferred'

A *Myrmica* worker taking nectar from an umbellifer flower.

species (usually grasses), where they may remove 100% (Pirk & Lopez de Casenave 2014). Again, this behaviour is not significantly reported in British ant species, although the heathland *Tetramorium caespitum* has been studied for its potential to spread invasive European thistle seeds in North America (Alba-Lynn & Henk 2010), and the common British *Myrmica rubra* is thought to be a significant disperser of the invasive European Greater Celandine *Chelidonium majas* there too (Prior *et al.* 2014). And seeds of the local (in Britain) Stinking Hellebore *Helleborus foetidus* are known to be dispersed by *Formica* and *Myrmica* in Spain (Garrido *et al.* 2002).

Incidentally, a recent study suggests that Brazilian harvester ants *Pogonomyrmex naegelii* tend to collect large seeds, 4–7mm across, from the close vicinity of the nest, but only smaller seeds (mostly around 2mm) when they travel more than about 5m away (Anjos *et al.* 2019). They instinctively (and very sensibly) realise the importance of travelling light.

Plant/ant relationships would seem to offer another rich area of research for the British field myrmecologist. Entomologists studying beetles, plant bugs, butterflies, moths, flies and bees spend an inordinate amount of time analysing the relationships between these insects and the plants they visit. Pollinators are high on everyone's list of 'useful' insects needing to be protected and conserved. There must be ample promise of reward for the ant specialist to work hand-in-hand with the field botanist.

Ants in the landscape

The recurrent theme of this book, that tiny organisms have massive effect because of their huge numbers, is nowhere more apparent than when we consider the influence ants can have on the landscape. It is immediately obvious if you take a stroll through the British countryside. Any area of rough grassland is punctuated by small or large hillocks; these are the ant-hill nests of the yellow meadow ant *Lasius flavus*, and they are far from transient soil heaps – they are potentially immortal. Worker ants are replaced on a week-by-week or month-by-month basis, but queens too can be replaced if they die of old age, disease or predation. As the ants burrow, the churning of the soil eventually has a visible effect on the landscape and an ecological effect on the other soil-dwelling organisms, both plant and animal.

Landscape architects

Though they are only 2–4mm long, the massed workers of *Lasius flavus* can shift shedloads of soil, and the distinctive domed nests seen on pastures and downs are long-lived, with individual hummocks recognisable for decades. A rough approximation of age can be made from calculating the above-ground size of the dome, and reckoning on one litre of soil heaved up per year. Pontin (2005) mentions domes of 180 litres from Staines Moor, on the Surrey/Berkshire border, which are more than likely nearly 200 years old. Blacker & Collingwood (2002) estimated 3 million *L. flavus* mounds in an extensive antscape across the Porton Ranges in Wiltshire with a combined population of 35 billion individuals.

This churning of soil particles must have a huge effect on the land, but just as earthworm activity is applauded by gardeners though rarely measured, so too any studies on subterranean ant activity are piecemeal and often anecdotal. Nevertheless, ants are often presented as major environmental engineers, shifting topsoil, controlling the numbers of their invertebrate prey or regulating their aphid herds. And that pre-biblical fascination with seed movement continues to hold the thrall of many soil ecologists.

True harvesters do not just remove seeds, but also collect bits of plant material from around the nests. This can be equivalent to the agricultural husbandry dished out by farmers clearing, ploughing, harrowing and hoeing the land, and ant effects can sometimes be seen on a landscape scale. North American *Pogonomyrmex* and European *Messor* harvesting ants are well known for creating large, mostly plant-free, circles around their nests, which when regularly spaced across the landscape are visible as polka-dot patterns in aerial photographs.

The enigmatic 'fairy circles' of Namibia are regularly spaced nearly plant-free circles of bare soil several metres across. They are visible in satellite photographs, and although they were once thought to be created by soil-dwelling termites, Picker *et al.* (2012) showed they were made by *Anoplolepis steingroeveri* ants. These are not, however, seed-harvesters, but attendants of subterranean leafhopper nymphs. The landscape effects were created by the ants excavating and aerating the root masses of the grasses on which the hopper nymphs were feeding – *Stipagrostis ciliata* on the periphery of the rings and *S. obtusa* in the middle of the discs. This tillage altered the drainage and moisture content of the soil, discouraging other plants from growing and thereby creating the distinctive flora-free circles.

BELOW AND OPPOSITE, TOP: The enigmatic 'fairy circles' of Namibia are visible in satellite images. They are created by *Anoplolepis* ants affecting the plant communities around their nests.

This is perhaps the one ant/plant effect that can also be readily observed in the British Isles, where the characteristic domes of yellow meadow ants *Lasius flavus* similarly punctuate the landscape in a regular array. They may not be quite big enough to register on satellite images, but are easily visible on aerial photos, and as you walk through a field of them they are striking enough features to make you stumble if you do not carefully watch your footing. The ant hills quite literally stand out from the meadowland surface, but they often also look a slightly different colour. This is nothing to do with the way the light hits them at a different angle or the shadows they may cast – it's because of the different plants that grow on top of the hills, compared to the matrix of the open field that surrounds them. On a recent walk in Knole Park, Sevenoaks, I was struck by the much darker green of the ant hills against the yellowing grass, grazed very short by the extensive deer herds. Here, the deep colour was mainly attributable to moss, but others I saw recently near Dorking in Surrey were crested by slightly richer tufts of more verdant grass. I assumed this was a result of the finer tilth on the ant hill, and possibly also its moisture content, but it might equally be the result of some subtle impact of the root aphids which *L. flavus* tends. I was not the first to notice this relationship.

Slightly fanciful interpretation of the 'cleared disk of the agricultural ant, with a central mound and seven roads', from *Life and Her Children* (Buckley 1880).

When I was a boy, rambling across the South Downs of Sussex on family outings, my botanist father would point out unusual plant associations, and he took delight in showing me Thyme-leaved Sandwort *Arenaria serpyllifolia* and Wall Speedwell *Veronica arvensis*, which both grew especially on the *Lasius flavus* ant hills but hardly at all on the surrounding hillsides. These were also two of several plants identified in the seminal papers by King (1977) as particularly occurring on ant hills, compared to the general matrix of the surrounding meadowland sward. Soil nutrients were not that different in the ant hills, the ants were not believed to be chewing the roots or selectively moving seeds, but the nest mounds consistently had a different flora. Both the sandwort and the speedwell are weak competitors and usually confined to disturbed soils where they escape shading by other more vigorous plants. The constant up-swell of spoil from the ants' digging helps keep the soil disturbed, and the prominence of the mound attracts rabbits which use the hummocks as lookout posts and latrines. Their grazing and digging further control rougher and more vigorous plants from taking over. Other plants that benefit from this dual ant/rabbit soil-disturbing activity include Common Mouse-ear *Cerastium fontanum*, Common Rockrose *Helianthemum nummularium* and Wild Thyme.

Despite their huge earth-moving abilities, ants remain very small and rather easily subdued by human activity. Even lack of activity can overcome some ants. Perhaps the saddest landscape

Mounds created by *Lasius flavus* in Knole Park, Sevenoaks, Kent, showing the evenly spaced array, and the darker green caused by mosses preferentially growing on the domes.

I can bring to mind is the barren floor of a straggling Hawthorn *Crataegus monogyna* and Sycamore secondary woodland in lowland England. Everywhere the ground is punctuated by dead ant hills, now merely rounded piles of inert soil with hardly a herb or a stalk of grass growing from them. Seventy years ago this would have been open wildflower-rich commonland, grazed at a low but effective level; however, changes in land use and the commercially non-profitable nature of small cattle herds or sheep flocks mean that the animals keeping the grassland open have gone; scrub invasion has passed to dense thicket and now spindly trees. This is not a habitat conducive to the warmth-loving *Lasius flavus* that would once have occurred here in their millions.

The low-growing Thyme-leaved Sandwort *Arenaria serpyllifolia* (left) and Wall Speedwell *Veronica arvensis* (right) both grow especially on the ant mounds of *Lasius flavus*.

Municipal architecture – the daunting and the decorative

In the British Isles, most ant-nest architecture tends to be small and rather vernacular. True, the ant hills of *Lasius flavus* and the nest mounds of *Formica* wood ants can be quite impressive, but they are relatively simple heap constructions. However, a brief world tour reveals some startling designs. The famously large and complex underground nests of leafcutters have frequently been revealed by using fast-setting plastic/latex, grout plaster, or liquid metal poured into them. These three-dimensional casts, several metres high and often as many across, can later be excavated, and mounted examples showing the various chambers and interconnecting tunnels make

Myrmecologist Walter Tschinkel gives scale to his cast of the nest of a North American harvester ant, *Pogonomyrmex badius*, made by pouring molten aluminium into it.

startling museum displays. The number of entrances is a good guide to the age and size of a nest – 1,000 in three years seems a benchmark number, by which time more than 1,000 chambers will have been hollowed out below (creating a 40-tonne spoil heap), about 400 of them active fungus gardens.

In India, small (4–6mm) harvester ants, *Pheidole sykesi*, fortify the nest entrance with raised concentric ring walls made of clay. The central entrance tube is about 7–10cm high, with gradually subsiding outer dykes, five to ten of them, extending to about 45cm in diameter. It is likely to be a defence against flooding, which can be very heavy in the region, and the nests are often built on a gentle downward slope. My first thought was whether J. R. R. Tolkien had seen them, inspiring his description of the concentrically ringed fortress city Minas Tirith in *The Lord of the Rings*.

Another, *Pheidole oxyops*, from Brazil, decorates the bare earth floor around the deep smooth entrance holes of its nests with bird feathers laid out like a carpet. This is thought to increase the number of

The Indian harvester ant *Pheidole sykesi* creates concentric ring walls of mud, thought to be a protection against flooding in heavy rains.

small prey insects that fall into the pitfall-trap nest openings (Gomes *et al.* 2019), even though the unrelated Asian *Diacamma rugosum* seems to benefit from moisture that accumulates around a similar feather decoration during the dry season.

In Afghanistan, various *Cataglyphis* species decorate their nest entrances with small stones. This is possibly where Pliny and Herodotus got the idea of gold-mining ants. The rocks warm up quickly in the sun, possibly transferring the heat into the nest soil to get the ants going first thing in the morning. *Pogonomyrmex montanus* does something similar in California (MacKay & MacKay 1985). Fossil hunters are known to scour these 'decoration zones' for small bones unearthed by the ants.

When Captain James Cook made landfall in Australia in 1768, it was not long before he and his on-board naturalist Joseph Banks discovered the football-sized nests of green ants, *Oecophylla smaragdina*, probably

In the Wadi Rum of Jordan, desert ants lay small stones near the entrance to the nest. The fact that they are arranged along one side suggests they are aligned to the sun.

Weaver ants, the Austro-Asian *Oecophylla smaragdina*, construct their spherical sewn-leaf nests first by grappling leaf edges together (above), then gluing them in place using silk from their larvae (right).

by walking into them. Rather frustratingly, many of the nests are built in small trees and shrubs at exactly head height. Just like *Formica* wood ants, green ants do not sting, but they bite ferociously and spray formic acid everywhere. The nests appear small and simple – bundles of leaves woven together to form spherical blobs starting as big as a clenched fist, but some reaching the size of a beachball. Structurally, though, they are marvels of engineering – mainly because they are built using the larvae as living glue-guns. First the leaves are hauled together by the combined brute strength of the ants working in animated scaffold chains, jaws fastened around waists. Scores of ants form a bridge, ratchet the large gaps closed, then work in zipper-like formation to precisely marry up the leaf edges. Once the leaves are touching, they are glued together. Banks marvelled at the combined strength of *Oecophylla*, which he tested by poking the overlain ant arrays with his finger only to see them scatter and the leaves spring back to their original positions. While one cohort of worker ants is grappling leaves, others go back to a neighbouring nest and collect some final-instar larvae. In the tender gripping jaws of their sisters, the grubs tense themselves and start to produce silk. Created as a sticky liquid when touched onto a leaf, the silk quickly hardens on contact with air, and through the stretching action as the ant daubs the larva to and fro, effectively zigzag-stitching the leaf edges together. Within minutes the silk is hardened and strong enough to maintain the bent leaves in their intended positions, so the manipulation chain-gang ants move off to grasp the next blade, and so the process continues.

Although most British ants make hidden and rather secretive nests, there is still wonder to be had in rolling over a stone to reveal the carefully crafted galleries made by *Lasius niger*. These often look as if they have been carved from the soil after the ant nouveau style, and the designs make highly instagrammable abstract images. And the paper carton nest of *L. fuliginosus* is truly remarkable in its construction. It looks as if the galleries have been hollowed out of the tree stump or log, but they have in fact been built up, using chewed wood fibres cemented together with saliva, and with a semi-symbiotic

BELOW:

Left: small *Lasius niger* nest exposed from under a large flat ceramic tile, showing the sculpted cavity walls and incipient brood chambers.

Right: carton nest of *L. fuliginosus* found in the floor void of a house in Guernsey. The curious householder had the presence of mind to photograph it and put out an interested plea on Twitter, rather than simply destroying it.

fungus, *Cladosporium myrmecophilum*, growing its mycelia through it to give it strength. There is no doubt that it is deliberately incorporated into the nest-carton matrix, and actively managed by the ants which add regurgitated aphid honeydew into the carton, presumably to feed the fungal hyphae. The fungus does not occur in the soil surrounding the nest, although plenty of other similar fungi are present there. Indeed, in experimental manipulations with the ant, antagonistic fungi, including several potential ant pathogens like *Metarhizium*, are inhibited by chemicals from the ants' bodies, but the nest-reinforcing fungus is actively encouraged (Brinker *et al.* 2018). This paper-making from chewed wood pulp is exactly how social wasps make their nest cartons and brood combs, but it has evolved separately in ants, and is known from several genera worldwide (Mayer & Vogelmayr 2009).

Ant conservation

Those hyperbole ant number and biomass statements from Chapter 1 are worth repeating here. Wilson (1987) suggested that close to one-half of all insect and one-third of all animal biomass on Earth is held in the bodies of ants and termites. In many parts of the world they are the dominant herbivores, granivores, seed dispersers and scavengers, and they have a huge impact on the rest of the invertebrate fauna. If ants are able to make these landscape-altering constructions above and below ground, with significant impact on other invertebrates or flora in the area, they become keystone species in the ecosystem. In Germany and Bulgaria, wood ants (*Formica rufa* and relatives) are protected species, as they are regarded as keystone predators, helping to keep forestry pests down. But where ants have ecological dominance in terms of great numbers, they do not have great numbers in terms of biodiversity. With only about 12,500 species worldwide, ants pale beside beetles (possibly 500,000 species), flies (maybe 250,000 but probably much more) and the rest of the Hymenoptera (at least 200,000). Nevertheless, there are plenty of rare ants that need conservation. Although it changes regularly, ants make up a significant proportion of the IUCN's red lists – 150 species according to Guénard *et al.* (2017), representing about 1% of known species. By definition rare insects are those that are rarely seen or found, so with ants this usually means tiny easily overlooked species, raising small colonies, in narrow niches, in obscure locations.

It is difficult (by which I mean impossible) to come up with any sensible suggestion for the rarest ant in the world. Bizarrely, *Cardiocondyla britteni* is as good (and as bad) as any because it is only known from a single worker found in butter beans in Manchester in 1919. However, it is patently not a British species, but likely an aberrant form of some Indian member of the genus. The relic *Aneuretus simoni* might be a contender. It's regarded as belonging to a primitive ant group, the only living member of the subfamily Aneuritinae, and known only from a few colonies in the rainforests of Sri Lanka. Likewise *Adetomyrma venatrix* (in the primitive Amblyponinae) is a small, pale, blind species known only from a few specimens found in the dry forests of Madagascar; the queen is, as yet, unknown.

In the British Isles, currently, 18 of our nominally 50 'British' ant species are scarce enough to warrant some sort of conservation concern status. These range from *Lasius brunneus*, which was traditionally regarded as a 'nationally scarce' (Falk 1991) ancient woodland species from central England, but which is now known to be quite widespread in old trees and may indeed be spreading, to *Formica pratensis*, the bog ant last seen in Dorset in the 1980s, which is now thought to be extinct on the British mainland. It is these uncommon

Lasius brunneus workers with grubs.

ants that actually tell us much more about the health of the environment than any number of common species. Whether any given species is scarce depends on so much, but for ants it very often comes down to suitable habitat, and the size and quality of that habitat. These two species tell a story.

When Norman Joy discovered *Lasius brunneus* in Berkshire, new to Britain, on 21 January 1923 (Donisthorpe 1923), it seemed that this ancient woodland species existed in a few relict locations in central England, at the very heart of the traditional broadleaved woodland systems that had dominated much of lowland Britain for centuries. It remained an uncommon species centred around Windsor and Reading, but by the 1970s it was known from broad swathes of central England, particularly the Thames basin and the Severn valley. Distribution maps continue to show it spreading – into Kent, East Anglia and Wales. It's often difficult to fully explain why ecological ranges change over time, but in the case of *L. brunneus* it seems clear to me that changes in woodland management are a good place to start. Before the Second World War, English woodlands were much more intensively managed for the wood they produced – not just construction timber, but wood for furniture including turned spindles, coppice poles, brushwood, wicker, firewood and charcoal. All this fizzled out during the second half of the 20th century, and woodlands began to deepen and darken. Without the regular coppicing, cutting and clearing regimes, they grew up into banks of large trees where previously a more open mosaic of glades, rides and copses prevailed. This has benefited an ant that likes to nest in mature trees in undisturbed woodland, although lepidopterists bemoan concomitant losses to open-woodland butterflies like the Duke of Burgundy *Hamearis lucina*, High Brown Fritillary *Fabriciana adippe* and Pearl-bordered Fritillary *Boloria euphrosyne*.

On the other hand, any ant of lowland heaths in England has had a hard time, getting harder, for several hundreds of years. Lowland heath is not worth anything to the agriculturally minded landowner, or indeed anyone; it was the desolate blasted heath of Shakespeare, and Thomas Hardy's Egdon Heath was rife with witchcraft and dark superstition. Consequently, heathland has been 'improved' (i.e. destroyed) so that crops can be planted, grazing animals can get fat, and dangerous-looking landscapes can be pleasantly domesticated. Between 1750 and 1980 95% of Dorset's heathland was systematically destroyed, mostly for agriculture (Stubbs 1982). No wonder *Formica pratensis* could not hang on. This is a trend that has continued as the endangered remnants of lowland heaths continue to be degraded by scrub invasion, urban encroachment and road building. Another victim of this decline is the narrow-headed ant *Formica exsecta*. Although there are still some colonies on Scottish moorland, the previous English strongholds in Hampshire and Dorset have disappeared, and during recent surveys only one English colony was detected – a small population in Devon. This species is now the focus of a concerted conservation project, run by Back from the Brink, to monitor and understand the insect's habitat requirements, and hopefully to feed this information into improved management of its existing and prospective localities in southern England. The aim is to discover exactly what the ant needs, and why it now only occurs (in England) in just a single Devon locality, and then to replicate the

necessary environmental conditions elsewhere and introduce the ant into new sites. A citizen-science project is running (even during lockdown) to get members of the public to record any wood ant nests in the area, sending photographs of nests and ants in the hope that some as yet unknown colonies of *F. exsecta* might be discovered.

Many declining ant species around the world are threatened by the arrival of those invasive and aggressive tramp ant species so derided for their attacks on humans, and agriculture. Although it has not yet reared its ugly head in the British Isles, in North America the imported fire ant *Solenopsis invicta* has been blamed for reducing local biodiversity, particularly in its effects on native ant species. This is a hotly debated topic. A recent almost pantomime exchange between research groups argued that it barely impacted local species (King & Tschinkel 2013a), oh yes it did (Stuble *et al.* 2013), oh no it didn't (King & Tschinkel 2013b). An editorial brokered a cease-fire, quoting the age-old plea of the scientist: 'There is clearly more research to be done' (Hill *et al.* 2013). Indeed invasive species do not always get it their own way. According to Mothapo & Wossler (2014), the Argentine ant *Linepithema humile*, common in Europe and now found out of doors in England, has not established in the

A nest from the lone English *Formica exsecta* population, photographed in south Devon in 2018 during the cross-organisation campaign 'Back from the Brink'. The nest is small and the plant fragments are finer than in other *Formica* species.

climatically suitable eastern escarpment of South Africa, because the local native big-headed ant *Pheidole megacephala* is a superior and more aggressive fighter, at least in those Lanchester power law battles involving equally sized armies.

Whichever vulnerable ant species need protection, the conservation of ants is different from most other insect-preservation strategies. While ants might be highly numerous at any given site, it is the number of colonies that represents the true status of a species, because it does not matter how many workers are roaming about, if a colony cannot survive to get to the point of producing sexual reproductives then it is doomed. For ants, conservation needs to be addressed to the superorganism, not the individual organism. If High Brown Fritillary conservation revolves around creating new flailed Bracken habitat for the mated females to lay their eggs so the caterpillars have the best chance of finding the right food-plant, in the right sunny, sheltered, niche, at least the butterflies can fly about looking for these spots. In this case the organism, with an annual life cycle, can react in that year to finding new habitat in which to lay its eggs. But the traditional idea of insects being able to bounce back because they lay plenty of eggs so populations can increase exponentially only works for those with annual life cycles. An ant nest is perennial, and although new nests are sometimes created from scratch, or by budding, the community of a species in any given locality is grounded on its existing nests, and their ability to keep producing new queens and males. A perennial colony (the superorganism nest) reacts far more slowly to a changing environment. It cannot get up and move (at least not far) to find new suitable habitat if circumstances change around it. For ants, it is the pre-existing nests that need to be conserved, as well as managing the surrounding habitat so that new nests can be built in the future. Although I may be stretching this a little, each ant nest is really a metaphorical rhinoceros. Rhinoceros conservation is about conserving the long-lived adult rhinos in an area, not just hoping

Although lacking antennae, which would make it too fragile, this sculpture of a wood ant, beside the nature trail at Joyden's Wood in Kent, is accurate and sympathetic – important in a very public woodland where deliberate disturbance of the many wood ant nests is a constant worry.

that any newborn rhino babies will somehow manage to grow up into adults several years from now.

Ant nests are particularly vulnerable to malicious human disturbance. I now feel terribly guilty about my childish stepping-stone games on the South Downs, even though they were half a century ago. I'm sure any countryside walker will have noticed that large wood ant nests are frequently stabbed with sticks, apparently only to goad the poor residents into a frenzied defence for the amusement of the uneducated. So, in order to educate their visitors, many nature reserves where wood ants occur have interpretation boards and other information on hand to try and inspire sympathy for the insects, rather than loathing and fear.

Looking after populations of even moderately widespread ants can also have important knock-on effects for the rest of the environment. Although many ants are fairly common, the myrmecophiles that live in the nests have another level of ecological complexity to contend with. When the Large Blue was declared extinct in Britain in 1979 (see page 254), there was an exasperated frustration among entomologists. The exasperation arose because they had just worked out why it was declining, but they got a seemingly very minor point wrong, and they were unable to stop its terminal fall. The butterfly's documented, inexorable, decline for more than 50 years had focused attention on the minutiae of its life and its intimate relationship with *Myrmica* ants, but had assumed ants were ants rather than working out which ant in particular. Even though landscape management was adjusted to ensure thyme and ants were assisted, the Large Blues kept vanishing. The realisation that turned things around came just a couple of years too late.

Making sure it's the right ant

When the tiny 1mg fourth-instar caterpillar of the Large Blue butterfly *Phengaris arion* drops from the thyme plant on which it has been nibbling, it can be picked up and adopted by any of five *Myrmica* species that regularly haunt chalk downlands in southern England – *M. rubra*, *M. ruginodis*, *M. scabrinodis*, *M. schencki* or *M. sabuleti*. But it turned out that it is not enough just to have any one of these five on the site. It was one species in particular that was needed. These five *Myrmica* species are closely enough related that under test conditions one species can be tricked into retrieving a grub taken from any of the others' nests if it is experimentally placed in front of a forager. The 'alien' *Myrmica* larva will be taken back to the nest, added to the host brood chamber by the worker, and reared as one of her own.

But this will happen only if the flow of nutrients into the nest is above optimum. Under times of food stress, the host ants close their tolerance window – it's all to do with the exact cocktail mix of those cuticular hydrocarbons by which one ant colony recognises and accepts (or not) members of another. When they are hungry, the nurse ants become highly sensitive to any differences in the chemical signatures of their charges, and anything that doesn't smell exactly right is singled out, killed, butchered and fed to the more correctly scented 'home' grubs. Those ant larvae from any other species, previously adopted without judgement, are now identified, slaughtered and nutritionally recycled.

Likewise, the larvae of the Large Blue emit a cloaking signal that will appease any of the five *Myrmica* species if the nest is in a state of benign sated calm. But in the lean times the chemicals they give off are more minutely analysed by the host ants. It turns out that Large Blue caterpillars have a scent closest to *M. sabuleti*. A caterpillar taken into a nest of one of the other four ants is initially on to a good thing, but in times of food hardship it is unlikely to survive the intense olfactory scrutiny, and it too will be eliminated and devoured.

Myrmica sabuleti is the least shade-tolerant of the chalk five, and only occurs on the hottest of chalk downland localities. Unfortunately changes in chalk downland management from the 1940s onwards have meant that it was *M. sabuleti* which was most often crowded out from the hillsides where the butterflies had once flown. Grazing, which had previously kept the sward short so that sunlight could readily warm the soil, had ceased to be commercially viable on limestone downlands, and farmers had removed their stock to more easily managed but *Myrmica*-unfriendly meadows down below. The grass on the hillsides had now grown longer, and even though thyme continued to grow and flower, the heat-loving *M. sabuleti* had been edged out – without them, the Large Blue caterpillars could not survive either.

Once this key fact was understood, it became imperative to return the hillsides to a grazed *sabuleti*-friendly state. It took a significant change in the grazing regime to bring back the grass sward to the right short length, allowing the warming rays of the sun to reach the

The Large Blue *Phengaris arion*: single adult (left), and a mating pair at a secret location in Gloucestershire (right).

soil and encourage the heat-loving ant to return. This done, in 1983, trial releases of a Swedish race of *Phengaris arion* (indistinguishable from the original British subspecies) showed that the butterfly could be successfully reintroduced.

The re-establishment of the Large Blue butterfly in England is one of the more heart-warming environmental stories. Starting in 1992, further releases were organised, and the butterfly was soon present on more than 30 restored chalk downland sites. By 2006 the annual census showed that over 10,000 butterflies were flying, the highest number recorded for 60 years (Thomas & Lewington 2010). Today the Large Blue is thoroughly reinvigorated in England, which now supports the largest known populations of this globally threatened butterfly (Thomas & Schönrogge 2019). The ant is also doing well.

A similar decline of the closely related European Alcon Blue *Phengaris alcon* has also been observed in former hay meadows in Hungary, where changes in grazing regime are affecting sward length and local *Myrmica* populations (Tartally *et al.* 2019). This butterfly lays its eggs on gentians, on which the larvae feed until they are adopted by *M. rubra*, *M. ruginodis*, *M. scabrinodis* or *M. schenki*. It is slightly different in that its caterpillars are not ant-brood predators, but are fed by ant trophallaxis. Nest infestation rates of only 4–5% mean that changes in ant abundance have a much greater effect on the butterfly. In this case, rather than grazing to get the sward to ant-friendly levels, burning the meadows appears to help.

293

Ants in a changing climate

The world is changing, and while fears around climate change may centre on melting icecaps, rising sea levels and flooding, there are likely to be other grave consequences including increased storm ferocity (and frequency), altered rainfall patterns, hotter summers, milder winters and who knows what else. Frightening times. But these are only the effects on humans. Other animals (and of course plants) are adapted to the very often narrow climate-controlled habitats that they inhabit, and they have been adapting to these for millions of years. The climate is now changing at a rate unprecedented since the end of the last ice age around 15,000 years ago. Inevitably there is a lot of supposition involved in trying to predict what the effects are likely to be, but insects, with their rapid generation times and mobile flying body forms, have proved valuable and interesting research tools here.

Many insect studies have focused on the potential mismatch between plants and the insects that either feed on them or pollinate them, if warmer temperatures mean faster bud-burst or earlier flowering times (DeLucia *et al.* 2012). This might impact pollinating bees and flies, or influence the general biodiversity of plant-feeding insects, but is perhaps less likely to have an effect on ants. However, increased atmospheric carbon dioxide typically increases the concentration of leaf carbohydrates (by increasing photosynthesis), but decreases the protein content, and this will have major knock-on effects for all herbivores, including the honeydew aphids tended by ants.

In temperate regions, future increasing temperatures (generally hotter summers and milder winters) might be expected to increase species diversity. A famous study of moth species diversity through a north–south gradient in Finland showed that for every 1°C increase in average summer temperature, an additional 93 Lepidoptera species occurred (Virtanen & Neuvonen 1999). This fits well with our ants' well-known higher diversity in the warmer southern portion of Britain, compared to the cooler north. Perhaps this bodes well for new and interesting species moving north, increasing biodiversity and providing novel subjects for citizen scientists to observe in their own back gardens. This is a false hope, though. It presupposes that the insects are able to make these movements easily – most are not. Certainly, a few common, widespread, adaptable and mobile species will take advantage of increased temperatures and spread north. These are the common or garden species that already occur regularly in gardens because they are generalists that can live in a broad variety of habitats, forage a wide range of readily available foods, and behaviourally have a habit for rapid dispersal. But most scarce insects are scarce because they are very sedentary, have precise habitat requirements and narrow food choices, and occur only in widely separated localities. If these uncommon species are to spread they will have to do so in a series of stepping-stone colonisation events over many generations. Unfortunately the 'natural' British landscape is already so fragmented that potential stepping stones for these habitat specialists are few and far between. The gaps between the stepping stones are too large and too dangerous, and mostly present unassailable barriers for a tiny vulnerable invertebrate, even one that can fly.

Narrowing our gaze back down to British ants, it seems likely that *Lasius niger* and *L. flavus*, already widespread and common throughout most of lowland Britain, will fill in the gaps in their British distributions as they colonise presently rather inhospitable northern hills and mountainsides. Both species produce large numbers of highly mobile flying queens which can found new nests on distant stepping stones, alone in the claustral style. Meanwhile, the red wood ant *Formica rufa* mostly expands its range by nest-budding. Its current hotspot, the hot dry sandy soils and mixed woodlands of the Kent and Sussex Weald, and the warm sunny heaths of Surrey, Hampshire and Dorset, are perfect, but even crossing the road to the next available habitat slot here is fraught with serious mortal peril. Any wider spread of this species can really only occur through that difficult and dangerous process of temporary nest parasitism, where mated queens enter the nests of other *Formica* species and gradually take over. And who knows how the interface of *F. fusca/F. lemani* will change? At present these two species are split along geographic lines – approximately the Aberystwyth to Flamborough diagonal (which I've just invented for the express purpose of improving knowledge of ant biogeography). These two species are not quite mutually exclusive, with a broad overlap (and potential hybrids) across this area of England and Wales. Intuitively *F. fusca* ought to move north, while *F. lemani* ought to retreat. But this demonstrates another danger of regional warming – any particularly cool-adapted species will soon run out of northern land onto which to move, and will find themselves edged out into the North Sea. They might climb up mountains, where an elevation increase of 300m lowers air temperature by 2–3°C, but Britain's mountains are not exactly lofty, so they will also soon run out of any hillside to climb. To some extent trying to predict how insects will react to global climate change is still poorly informed guesswork; at the moment we simply do not know enough about most insects' precise habitat needs to work out if and how climatological alterations will affect them. And for ants in particular there are many confusing and compounding variables. For example, we know little about queen or larval physiology compared to that of adult workers, it's unclear how winter temperature changes might affect hibernating colonies, and heat/desiccation effects are poorly studied.

Having said this, we do know a lot about *some* ants – they are the widespread, adaptable and mobile species that are very likely to spread in the wake of global warming. They were spreading ahead of it – those invasive tramp species already well known to be causing ecological upheaval in their new regions of conquest. These are already internationally successful species, mainly because they are highly adaptable, non-specialist omnivores. There are about 200 ant species that have successfully established themselves outside of their natural 'native' range, but there is a top 19 most invasive given in the extensive recent review by Bertelsmeier *et al.* (2016), topped by the usual suspects – the red fire ant *Solenopsis invicta* and the Argentine ant *Linepithema humile*. Though the pharaoh ant *Monomorium pharaonis* is also invasive, it is primarily an indoor pest species throughout much of the world, so is insulated from outdoor climate and weather changes.

Climate change itself is not necessarily expected to increase the number of ant invasions; these, after all, are usually on the back of human-trafficked goods such as pot plants (orchids, bromeliads, fruit trees and acacias in North America), but it is thought likely to increase the successful colonisation and spread once they have arrived. It is also likely to extend the range through which an invasive species can invade, as those once-limiting temperature barriers are moved. *Solenopsis invicta* is expected to move significantly further north into New England and the Great Lakes area of the United States as temperatures increase.

Meanwhile, in the tropics, things are likely to get harder for ants. Counterintuitive as this might seem for these heat-loving insects, species living in hot environments are already working near the top end of their optimum heat performance window. For every insect species there is a relatively narrow temperature range through which it can function and survive. If it gets too cold its metabolism slows to the point where it cannot physically move or even metabolise, and a minimum of 10°C is often quoted for most ants. There will also be a low-temperature threshold below which an ant colony cannot get enough foraging momentum to produce sexuals – the 20°C threshold mentioned in Chapter 6 for some *Myrmica* species in Scotland, for example. As regional temperatures increase, the ants become more active, forage better, build bigger, and produce queens and males faster – this is why Britain's ant fauna is concentrated in warm south-eastern England, and why ants in general thrive in the tropics.

As temperature increases, metabolic and developmental processes also speed up; *Linepithema* eggs hatch in 58 days at 18°C, but only take a quarter of that time (15 days) at 30°C. At some point there is an optimum, a best equilibrium between foraging ability and reproductive success. But as temperature continues to increase there comes a point at which physiological processes cease to function properly – the 'upper thermal tolerance'; it is just physically too hot to go on. The biochemical reactions that keep the machine of the body running smoothly start to go awry as the complex enzymes, fundamental to all living organisms, change their three-dimensional shapes, profoundly altering their activities or disabling them completely. Most tropical ants are already running hot near the top end of their physiological capability; if climate change brings warming to these tropics, even a couple of degrees may be too much for the ants. Also, most tropical ants are arboreal, so the usual colony thermoregulation allowed by moving brood up and down in the soil heap is not available to them. A thorough review of how global warming might impinge on ants is given by Chick *et al.* (2017). Much depends on behavioural plasticity and whether the ants will be able to adapt behaviourally, if they cannot adapt physiologically.

The outcome of climate warming is likely to be loss of biodiversity, not just of ants, not just in the British Isles, but of all organismal groups, across all of the world. By observing ants, myrmecologists are not just close-focus studying a small number of standalone insect species. They are researching keystone animals that have landscape-wide effects, and which interact with a huge range of other organisms in a broad ecological web. Ecology is the interconnectedness of every living thing on the planet. Ants are a significant part of that connectivity web. Studying ants (and other insects) allows biologists unprecedented access to an understanding of life on Earth, and how it is changing right now, right in front of our eyes.

How to study ants

Though ants are seemingly everywhere, it can still take quite a lot of searching to find them – at least the unusual ones. As with all plants and animals, it soon becomes clear that the world is dominated by a small number of virtually ubiquitous species. These make useful subjects for broad behavioural studies like monitoring mating flights, observing colonial interactions, watching trophallaxis in action, or mapping genes across geographical zones. But these common or garden species, common in gardens and indeed almost any habitat anywhere, don't actually tell us very much about the health of the environment. A closer look at the other species is required.

First find your ant

In the British Isles the black pavement ant *Lasius niger* and the yellow meadow ant *L. flavus* are everywhere in pavements and meadows respectively, and various other common and widespread habitats like parks, gardens, hedgerows and verges in between. But slotted in among colonies of these are all the other ant species. Some are locally abundant, like the wood ant *Formica rufa*; some are widely scattered, like the secretive *Temnothorax nylanderi*; some barely cling on to a toehold on the south coast, like *Solenopsis fugax*; and some verge on the mythical, like *Myrmica lonae*. It is the combination of widespread and scarce species that makes up the full diversity of British species. Those that are adaptable and flexible can occur almost anywhere. No matter what we do to the environment (be it tarmac or Astroturf or close-mown utility playing fields), these ants will be there, so they tell us very little about how well (or not) we are managing that environment. On the other hand, it is the less common species, those with more particular habitat requirements, occupying narrower

OPPOSITE PAGE:

You have to get close to study ants, but be careful if they do this – a *Formica rufa* worker prepares to squirt formic acid against a perceived threat.

ecological niches, that can tell us by their presence (or by their absence) something about the health of the countryside.

This is where field entomologists come into their own. I hesitate to use the word 'amateur' with its tainted suggestion of the casual, unprofessional, inexpert hobbyist. Even today, when there is a veritable industry concerned with environmental protection, land management and wildlife conservation, most of the actual wildlife recording is done by interested private individuals, in their own time, and at their own expense. They may not have the honorific title associated with a museum department, or tenure at an academic institution, yet this body of scientific citizens has in its midst national experts and world authorities. Just by picking up this book and reading it you are part of that citizenry. Welcome.

Traditionally it has been the task of the field entomologist to record what species occur where, and to make comparisons between different localities, and different habitats – to map out geographic distributions (expansions and declines), observe local behavioural and ecological phenomena, and to help build up a bigger picture of the whys and wherefores of insect biology. Ants are no exception, and although the Aculeata (bees, wasps and ants together) have garnered greater support and interest in the last 25 years, this effort still pales into insignificance compared with the work done on plants and birds,

OPPOSITE: A hand-coloured plate from Curtis's *British Entomology* (1823–1840), showing only the male of *Myrmecina graminicola* – since this was all Curtis had to hand when it was found, new to Britain, at Puckaster Cove in the Isle of Wight.

THE ANT-HILL.

If you are going to stop and observe an ant hill at close quarters, it pays to tuck your breeches into the top of your socks, to stop the little blighters crawling up your legs. From the quaint *Fairy Frisket; or Peeps at Insect Life* (Tucker 1889).

butterflies and dragonflies. There is still much work to be done, and even though there are only about 60 species known in the British Isles (about 50 'natives' and 10 introductions), ants can still make a valid and valued contribution to our knowledge of the countryside. So, finding that ant …

The ants are there to be found. The trademark stooping stance of the insect photographer can be adopted by anyone, camera in hand or not. Simply by patiently staring at leaves, stems, flowers, tree trunks and the bare ground, ants can be seen. Through the camera viewfinder a magnified world not ordinarily seen by humans is revealed, and close observation is the first stage of a voracious curiosity that can soon consume your every moment when out walking in the countryside.

Snapping ants: photographing portraits and behaviour

Though ants are small, camera technology is now good enough to put insect photography into the easy reach of almost anyone. Every year new cameras and lenses appear, so any recommendation of brands, models, gadgets and gizmos will be out of date before this book is published. So rather than prescribe any fixed set-ups, the following is merely a guide to techniques.

Modern single lens reflex cameras come with detachable lenses and the ability to take a specially designed macro lens, or extension tubes slotted in between lens and camera body. Alternatively,

Typical pose of the insect photographer, on hands and knees on the ground. Penny Metal, many of whose photos appear in this book, uses a short macro lens with ambient light.

Lasius niger lured to a droplet of sweet liquor on a brick, to be photographed at leisure.

a supplementary lens can be screwed in to the end of the normal lens; this is a bit like adding a small magnifying glass to the camera. These methods effectively allow very close focusing on an object only a few centimetres away, with concomitant magnification to life size or above. They maintain through-the-lens light metering, but as magnifications increase any auto-focus mechanism may need to be disabled, as the camera often cannot cope with even a slight movement at such close quarters. Whether to use flash or ambient light is a personal choice, made by balancing the amount of paraphernalia needed against ease of use. At higher magnifications artificial light improves the depth of the focus field; a ring-flash, twinned flashes, or some sort of translucent diffuser shield helps avoid any stark shadows. Stacking many images shot together in quick succession, using a computer to process the files, is an up-and-coming technique. Consecutive shots, each with a slightly different part of the insect in focus, are merged to give a composite image. Very high-resolution pictures of very small insects are possible using a complex rail-aligned gantry in the studio, but stacking of hand-held pictures can work, and may be increasingly possible in the near future as technology advances.

Whatever the mechanics and aesthetics of taking the picture, much of natural history photography relies on the cooperation of the subject, and photographing insects is fraught with all sorts of difficulties. Getting close enough without disturbing them is one of the biggest obstacles. At least most ants cannot fly away, but they are fast on their feet and are not impressed by the looming approach of even the most benign smiling camera-wielding entomologist. A good trick is to photograph insects when they are occupied – feeding, mating or grooming – and this is especially pertinent when photographing ants. Standing back to observe, then a slow approach, taking care to avoid your shadow falling over them, helps. Ants may rush about headlong on the trail, but occupied en masse with a large prey item, tending their aphid herds, or stopping to exchange food by trophallaxis often allows a few moments for the photographer to get the shot. Baiting can work well with ants – simply apply a blob of honey, yogurt, meat paste or sugar water on or near the trail and they will soon be lining up to have their pictures taken.

Patience is the photographer's secret weapon. Accidentally disturbing a butterfly, bee or fly from a flower might be remedied by standing very still for a few moments and waiting for the insect to settle again nearby. Ants are also creatures of habit, so if they run away at top speed, wait and see whether they (or their nest-mates) return to carry on the behaviour shortly. Or come back later, or tomorrow, to catch them at it again if necessary.

To get over camera-shake, bird photographers often use a tripod, but this makes the whole set-up unwieldy when trying to get to within a few centimetres of an insect. When photographing an ant on solid ground, it is usually possible to use the non-shutter-firing hand as a mini-tripod, resting your fingers on the ground and using the back of your hand to support the camera lens to give a stable image. At this close to your target, focusing is simply a question of moving the camera back and forth a few millimetres to get the best crisp picture. Be prepared to take a series of photos. One of them will be in focus, or show the best lighting, or capture that exact moment when the ant is doing that something you were trying to catch.

Ironically, warm sunny weather, when insects and entomologists are most happy to be about, often makes it more difficult to take photos, because the insects are so active and agile. In cool, overcast (but not wet) conditions they will be more sluggish. Similarly, early-morning photographic forays can be more rewarding, when your subjects are still warming up.

Whatever your target group, whatever your equipment, and whatever your aesthetic preferences, if you want to photograph ants (or indeed any insects) be prepared to spend a lot of time scrabbling about in the dirt on all fours, and to explain your antics to curious passers-by.

Collecting up ants

Because many ants are rather robust little creatures, some individuals can be carefully picked up between finger and thumb and deposited into a container for closer examination. No British ant species is powerful enough to puncture fingertip skin with mandible or sting, although large wood ants have a great ability to seek out that delicate fold of skin between forefinger and thumb and take a firm nip. I always feel great admiration for the diminutive creature attacking me, the monster, like this, but I know some people get highly agitated at the thought of being bitten by any insect.

Even without picking them up, an ant can often be coaxed into the tube. I normally carry a tin containing a variety of glass tubes from 11mm to 20mm in diameter. Some myrmecologists advocate fine-tipped tweezers, but I find them rather awkward and inconvenient. For the tiniest *Monomorium* and *Temnothorax* species, I sometimes pluck a small piece of grass, lick it slightly and use the stickiness of the

The correct use of the pooter.

spittle to gently lift up the ant before placing it and grass into the glass vial. The pooter (sometimes called an aspirator) is also useful at this juncture. Named after its inventor William Poos (1891–1987), this comprises a glass container with two flexible rubber tubes and comes in various designs (Leather 2015); it should not, however, be confused with the amusing fart toys of the same name marketed in North America. Sucking on the long tube creates a vacuum, and small delicate insects are siphoned up through the short inlet tube; they are held in the main body of the pooter, prevented from being gulped up into the mouth by a thin gauze barrier. Be careful with wood ants, though – they are apt to eject a cloud of formic acid, and this easily passes the barrier to leave a strong taste in the mouth. Maintenance of your pooter is paramount, to

avoid the frequent tales of losing the mesh barrier, thus creating an elaborate straw to inhale live insects, or the realisation that 'I'd been using the pooter for three weeks before I discovered there were two dead mealworms lodged inside the inhaling tube' (Naomi Harvey, very personal communication). Many non-entomologists have heard of the pooter and are curious to see its demonstration first-hand in the field – always a good opportunity for a bit of entomological outreach. It can be misused, though – on the week-long ecology field-trip to Ashdown Forest during my first year at university, we discovered in the heather a discarded pooter from a previous visit, and there was no doubt whatsoever that it had been used as some sort of experimental hubble-bubble hashish hookah pipe. Students, eh?

My standard entomological tool is the sweep net, with which I trawl through the herbage and low-hanging branches with slightly comical abandon. This is a low-tech, but admirably flexible, all-round method and suits almost any locality on any occasion. I like to think that if Charles Darwin, Alfred Russel Wallace or Horace Donisthorpe stepped out of a time warp in front of me they would know immediately what I was doing, and why. Members of the public also guess that I am after insects, and are often fascinated to find

Using a sweep net – so easy a child can do it.

out that there is more to entomology than catching butterflies. A few swings of the net through the long grass, and there is plenty to see. Simply peer in to find what has been knocked in there. It's usually massed workers of *Lasius* or *Formica*, but the occasional winged *Myrmecina graminicola* appears. A beating tray, effectively a piece of white sheeting spread taught between radiating spokes, can also be used under tree branches smartly smacked with a stick to knock down insects, often *Lasius brunneus* or *L. fuliginosus*. I mostly use a pale umbrella upturned; it's smaller than a purpose-built beating tray, but it folds up more neatly for my backpack – oh, and it can also be used to keep off the rain.

'The umbrella and its mode of use', from the introductory entomological text *The Insect Book* (Howard 1904).

Another simple technique for finding ants is visual searching and fingertip grubbing. Tree trunks and bare soil can be watched for typical ant movements. Most British ant species appear to be diurnal, active only during the day (this is most likely to be related to

Temnothorax nylanderi nest found inside the layers of the Cramp-ball fungus *Daldinia concentrica*, broken off a fallen Ash *Fraxinus excelsior* trunk.

Ants will sometimes visit animal dung, either after small prey or for moisture. Here is a *Formica rufa* worker under a New Forest pony dropping.

air temperature), but searching at night by torchlight is recommended by some researchers in tropical and subtropical regions. Logs and rocks can be rolled over; this is usually the best way to find *Myrmica* nests. Pieces of loose tree bark can be prised away, and fungal growths can be examined, for the small secret nests of small secretive *Lepthorax* and *Temnothorax* ants (they will also take up residence in empty snail shells, hollow acorns, fungi and rotten conkers). Few books on dung insects mention ants, but they do visit, either after small prey or for moisture. Please wash your hands after dissecting a pat, though.

Low-growing herbage can be rootled through, grass root-thatch can be manually separated, leaf litter can be sieved over a plastic sheet. This is the type of behaviour most likely to raise concerns from passing members of the public wary at seeing you, on all fours, backside in the air, ripping at the base of grass tussocks with your bare hands. I find a cheery wave and a puff on the pooter sends them on their way; or I engage them in polite and earnest conversation about the importance of insect study when it comes to nature conservation. And generally, people are interested to know what I'm up to. Despite the fact that the insects I'm studying are only a few millimetres long, there is an increasing awareness of their importance, and of their interconnectedness in the wider ecology. Everyone knows an ant when they see one, they regularly appear on nature reserve information boards and displays, and it is heartening to see them used to generate genuine concern for the environment.

Other techniques for finding ants include baiting and pitfall-trapping. These are longer-term studies that require return visits to a site to assess the catch. Well-tested baits include tuna or sardines in oil, and peanut butter; these can be smeared on tree trunks, fence posts and rocks or left out in the open on a piece of wood or slate. Once the bait is discovered, the trail of foragers can be followed back to the nest and recruitment/trophallaxis behaviours can be

Quite literally, a wood ant. Public engagement at the National Trust's Avebury Manor in Wiltshire.

observed. Pitfall-trapping involves setting small containers, such as plastic disposable drinking cups, down into the ground so that the rims are flush with the soil surface. They trap small surface-dwelling invertebrates which fall in and cannot climb up the smooth sides. Again, these need to be visited regularly to empty out the catch. Dry traps should be visited the next day, otherwise the many invertebrates that fall in, including beetles and centipedes, start to eat each other. Wet traps can be loaded with a few centimetres of water and a drop or two of detergent (washing-up liquid). This does mean that the trapped insects are all killed, but this also preserves them for a short period and, since it avoids in-trap predation, the specimen counts are more robust in statistical analyses. Wet traps can be left for several days before a follow-up visit. Other wet-trap variants include using concentrated salt solution, strong white pickling vinegar, or antifreeze (ethylene glycol) as longer-term preservatives, but care is needed because ethylene glycol is poisonous to Badgers, Foxes *Vulpes vulpes* and dogs, which are attracted by the slightly sweet smell and are known to dig up the traps to eat the contents, with potentially dire consequences.

For the keenest myrmecologist, a suction sampler might prove useful. This is a converted two-stroke garden blower-vac, with a stiff canvas bag firmly attached over the intake tube. Once the device is going it hoovers up invertebrates from short herbage and they become trapped by the moving air in the bag. When it is stopped, the captured insects can then be tipped out onto a plastic sheet to be examined. This is a good way of finding tiny non-flying insects when the herbage is too short for the sweep net. When I bought one and tested it on the irregular patchy lawn of my suburban garden, I sucked up many specimens of *Myrmecina graminicola* and *Ponera coarctata*, neither of which species I had previously found there (Jones 2004). Suction samplers are also useful if you are trying to make population counts, although a pooter and a good amount of suck might also suffice.

Using a suction sampler on the *Sedum* 'living roof' of the Canary Wharf shopping complex in London, 2002. *Lasius niger* was the dominant ant species here.

Counting ants

Seeing a trail of ants might give the impression that they are easy to count, and so they are as they run by along the forage route. But relating this simple count to the actual number of ants in an entire colony is more tricky – you can't count them all out, then count them all back in again. Instead, an estimate of population size can be gained by the mark–recapture technique. This means collecting and marking some live specimens, letting them go back into the crowd, making another capture sample sweep a day or so later, and counting how many of the marked individuals turn up in that second sample. This is a frequently used method, and the simple mathematics to give an answer are based on:

$$N = Mn/m$$

where N is the calculated total colony size, M is the number of marked individuals released, n is the total number of specimens collected on the second count, and m is the number of marked individuals caught up in the second sweep. Simple.

Unfortunately, it is never as easy as it sounds. First, how do you visibly mark an ant without damaging it or endangering it by making it stand out more for predators, or smell different so it is likely to be rejected by its nest-mates? Butterflies and large beetles have been marked using permanent marker pens or blobs of coloured nail varnish, or by having small tabs of coloured paper glued to them. All these methods are likely to have a negative effect on the individual ant. They also involve quite a lot of insect-wrangling; this is never going to work with tiny ants.

There are several techniques, but each comes with a level of technicality that the average field entomologist might find awkward. The simplest involves diluting a fluorescent dye or ink in ether, then using a disposable perfume atomiser to spray onto the first sample of specimens. To make the numbers more statistically robust, sample sizes of about 300 individuals should be collected. This is where the vacuum collector comes into its own. A few minutes of hosing up and 300 ants are easily accumulated. A few squirts from the atomiser and they should all be tagged – but don't overdo it or they'll all be gassed by the solvent. Allow a few minutes for the ether to dissipate and the ants can be set back among their sisters to resume foraging duties. At the next vacuuming sample session an ultraviolet black-light can be held over them to see which ants fluoresce, and they can be counted against the non-marked nest-mates and the calculation can be done.

Other techniques to mark the first sample of ants include spraying with aerosol tins of paint (from about 50–100cm so they do not become clogged), slowing the ant by exposure to carbon dioxide and carefully dobbing a blob of quick-drying enamel paint onto the gaster (larger species only), or dousing them in a spray containing a radioactive ^{32}P tracer (OK, perhaps not so easy for the citizen scientist).

In all of these methods, one fundamental problem is that the samples are only ever of foragers, which represent only a portion of the colony's occupants. Whether they represent 95% or 25% of the colony might depend on the species, the age of the colony, the time of day, the season, whether food availability is high or low – so in the end the calculated figure can only be claimed to be an estimate of the forager cohort size, rather than of the entire nest. Well, at least it's a start.

Once found, what next?

One of the key pieces of field equipment for any entomologist is a simple hand lens. Don't be fooled by the large Sherlock Holmes grasping-handle type magnifier – it's of no use other than to read the classified columns of the newspaper down at your club. Small folding pocket lenses (glass, not plastic though) can be bought for just a few pounds at wildlife centres, optical stores, hobby shops, jewellery and watchmaker suppliers. They are robust, good quality, and easy to use. Entomologists (and botanists) are easily identifiable in the field because they usually wear a lens on a ribbon or lanyard around the neck for convenience. Low magnification ×10 is perfectly good for almost all field work and will allow many identifications in the field. Using a hand lens will also aid any decision to collect a specimen for confirmation under the microscope.

This brings us to the all-time paradox faced by entomologists, and the act that might yet alienate them from a curious public looking on from the sidelines – why do entomologists have to kill the objects of their study? The answer is that nobody needs to kill an insect to study entomology, but if you do decide to make a reference collection of insects, your knowledge and contribution will be profoundly deeper and longer-lasting. The arguments are simple. Taking a few sample specimens will in no way exhaust or damage the colony. Indeed I have often made the claim that taking specimens of almost any insect will not harm populations, or the environment. Their huge numbers and prodigious fecundity, combined with their small size and secretive lifestyles, mean that entomologists only see a tiny fraction of them anyway. And there are far worse dangers out there – mortality from predation by each other, spiders and birds is huge, and many other human activities have a much more drastic effect on them. Indeed, my back-of-the-envelope calculations, based on an easy underestimate of 1,000 insects (many of them ants) per square metre, suggest that the building of the 2012 London Olympic Park (2.5 square kilometres) killed roughly 2.5 billion insects, more than have been collected by all the entomologists in the world, who have ever lived. And given that people so easily reach for the can of pesticide spray if they find insects indoors, it seems hypocritical to chastise entomologists for collecting a few field specimens. On the other hand, taking the specimens will provide an opportunity for them to be examined carefully and comprehensively under a stereomicroscope, using the sometimes complex identification keys to confirm identity and validate, to the highest degree, the scientific worth of a field record or observation. True, some insect species can be identified in the wild by using that cheap hand lens, but this expertise needs to be gained, and I suggest it needs to be gained first by making a collection of specimens, and checking them with the microscope. Specimens that form part of a personal reference collection, or eventually become integrated into national or local museum collections, have the added advantage that they can be examined and re-examined again, and again, perhaps for centuries to come. As taxonomic revisions adjust our interpretation of which species is which, only a physical specimen can be looked at in the light of new understanding, and much of what we know about recent species arrangements is based wholly on looking at dead specimens in private or museum collections. When *Stenamma debile* was recorded in Britain, all previous records of *S. westwoodi* had to be disregarded unless an actual specimen could be re-examined and redetermined.

The decision to collect insects to examine under the microscope is a personal one, decided by each of us as the opportunity arises in the field, and is a complex scientific, moral and ethical calculation based on whether a specimen is necessary for correct identification, how likely that identification will be, how important the record might be, and whether a specimen might improve a reference collection or future knowledge of the species. Gone are the days of butterfly-collectors vaingloriously stocking their cabinets with prize possessions to be gloated over in secret or shown off privately to zealous fellow cognoscenti. Today the reference collection is a workaday scientific tool to aid identification, and ultimately to aid understanding and knowledge of insects.

One important thing about ants is that they are small, easily prepared for the reference collection, and easily stored in a relatively small space – no need for the expensive mahogany cabinets of the lepidopterist. The other good news is that entomology, generally, is a cheap and convenient science to study. This has long made it one of the most egalitarian of sciences, since anyone can find insects, anywhere, and they can do so easily and at low cost. Most equipment and materials can be purchased cheaply, repurposed from common domestic products, or cobbled together from easily available sources. The only major outlay may be an inexpensive stereomicroscope. A £100 budget model should get anyone started (far less than most mobile phones), though eye-wateringly expensive top-end professional models can be the price of a small car. Only low magnification is required, say ×10 or ×20. This is similar to a hand lens, but the brighter clarity of the optics, being able to see the stationary specimen in clear stereo vision, and the ability to carefully manoeuvre the insect around for different views make the microscope an invaluable tool. Zoom optics are slightly more expensive, but different magnifications can be achieved on the cheaper models either by having swivelling turret-mounted objectives or by swapping eye-pieces. Some microscopes have built-in lighting, but almost any budget scope will benefit from the additional use of a small desk-top anglepoise lamp.

The aim of collecting ants (or indeed any insect group) is to have a series of labelled and ordered specimens that can be easily manoeuvred for examination and comparison without damaging them. Insects, with that hard external skeleton of chitin, are easy to preserve – they just need to be dried out, after which they become brittle, but they do not rot. No artistic taxidermy skills are required. The two standard practices for preserving insects are pinning and carding. This is another personal choice based on the practicality of dealing with different-sized specimens.

First kill the ants in a small glass bottle or jar (not plastic, which is attacked by the solvent) with a twist of tissue paper doused in ethyl acetate – this is sold as 'killing fluid' by entomological suppliers, but is also the major component of acetone-free nail varnish remover, available from almost any pharmacy or supermarket. After about 10 minutes, the insects are dead, but still flexible. Some entomologists routinely drop specimens into 80% alcohol when out in the field. This immediately kills the insects, and preserves them almost indefinitely. They can then be examined in the alcohol under the microscope in a shallow petri dish or dried and mounted for the reference collection. Neat ethyl alcohol is sometimes difficult to obtain since excise duties mean a special government licence is required to buy it tax-free. When I duly filled in the Home Office form, using the usual black biro as requested, I suggested my annual use would be <5 litres per year. This seems to have been misread, and when my licence arrived it turned out that I was granted permission to buy 75 litres. Organising to buy large quantities of industrial chemicals is the sort of thing that would set alarm bells ringing nowadays. Various over-the-counter preparations like rubbing alcohol work well enough. When I worked with a citizen-science project in Regent's Park in 2017, the specially ordered alcohol had not arrived in time, so two bottles of cheap vodka were purchased instead. Another method, one advocated by BWARS to humanely kill bees and wasps, is to place the insects in the tubes in the freezer overnight. This induces them to go into roost 'sleep' mode, then hibernation torpor, before painlessly killing them.

Large winged queens, and even small winged males, can be pinned like conventional insect specimens (see page 316), but typically small workers should really be glued to a piece of card. Acid-free four-sheet Bristol board water-colourist card is the averred choice of the museum professional, but any thick smooth white card (250–350g/m²) from a stationer or art shop, or even bits cut from old Christmas cards, will do fine. Various professional-quality glues marketed as gum Arabic or gum tragacanth are mentioned in old textbooks, and are still available from specialist suppliers at relatively low cost, but at a pinch I've used thick wallpaper paste or slightly watered down PVA craft glue. You don't need much. Try and brush out the legs of the upturned ant using a small dry water-colour paintbrush, before placing it carefully, right side up, onto the glued patch on a piece of card. Next try and neatly spreadeagle the limbs and antennae using a small needle stuck backwards into the rubber eraser end of a pencil. For the very smallest specimens I use a thin stainless-steel headless pin stuck into the end of a matchstick, but various Victorian manuals suggest a stiff horse bristle or a cat's whisker – whatever works. I usually do this under the lowest power of the microscope (×5–×10), but adjustable desk-top reading magnifiers or 'linen prover' lenses will do just as well. This 'setting' helps make all the relevant body characters visible when trying to identify the ant later. Having all the ant specimens similarly neatly splayed, like the pictures in the identification guides, makes it much easier to appreciate some of the more subtle comparative characters when examining two specimens side by side.

When the glue has dried, after an hour or two, the card can be neatly trimmed to a small rectangle around the specimen. A pin stuck through the end of the card allows the whole shebang to be picked up and moved, angled under the microscope objective, or firmly stuck into the airtight storage box in which the ant collection is being arranged. The card serves to prevent limbs and antennae being knocked and broken; the pin acts as a holding handle, reduces jarring on the specimen, and can take the all-important data labels – the more the merrier (see page 316).

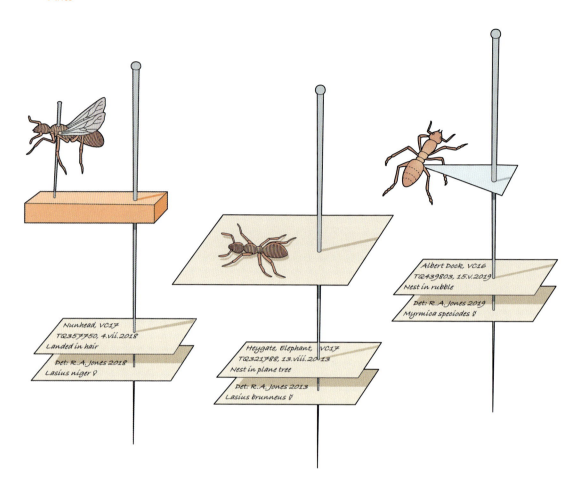

Nunhead, VC17
TQ357750, 4.vii.2018
Landed in hair

Det: R.A.Jones 2018
Lasius niger ♀

Heygate, Elephant, VC17
TQ321788, 13.viii.2013
Nest in plane tree

Det: R.A.Jones 2013
Lasius brunneus ♀

Albert Dock, VC16
TQ439803, 15.v.2019
Nest in rubble

Det: R.A.Jones 2019
Myrmica specioides ♀

The standard methods of mounting ant specimens: pinning, carding, pointing.

Alternatives to carding are pinning and pointing. Most British worker ants are too small to pin in the conventional way, as dipterists or lepidopterists might pin flies and moths, for example. However, winged queens and males are sometimes large enough. The industrialisation of pin manufacture during the 18th century apparently revolutionised entomology, making standard uniform pins available at low cost. Most haberdashery pins are too large and thick, and the metal alloys corrode over time as they react with the insect body, but modern stainless-steel pins come in a variety of thicknesses and lengths, and are available cheaply from specialist suppliers (who usually also make longer versions of them for acupuncturists). A slender pin (0.1mm diameter is my usual choice) through the centre of the thorax is then embedded in a mounting strip of cut Plastazote (a flexible and resilient expanded polythene foam that grips the pin well; again, cheap to source) and a larger stouter pin is passed through the other end. This large pin is the usual manipulation handle by which the specimen can be picked up and examined, and it should take the same detailed data labels. Although the delicate legs, antennae and wings of the ant are exposed and prone to damage by being knocked, the pinned mounting strip acts like a suspension/damper system to avoid unnecessary jarring – nevertheless delicate care should always be used when picking up and moving all set insect specimens.

Pointing is a technique by which a slender triangle of that thick card is anointed with a droplet of glue at the very tip, and the underside of the ant's alitrunk placed onto it. A large pin through the base of the triangle again takes the data labels. This is a useful alternative for very small specimens and allows a better view of most underside characters, some of which can be obscured if the ant is glued out flat with the standard carding format. It does leave the insect slightly vulnerable to knocks, though.

Founded in 1787, wire merchant D. F. Tayler made needles and haberdashery pins before branching out into the entomological market. These are from about the 1940s.

Never argue with the data

It is impossible to overstate the importance of data labels attached to specimens. A museum specimen without data is scientifically useless, but a well-labelled display destroyed by museum beetle or other pests still retains its information, even if the original insects are lost. A label should contain the absolute minimum of location where the insect was found, and the date. Ideally it should list locality name, geographic region, six- or eight-figure Ordnance Survey grid reference, date, name of collector, details of habitat, how the insect was collected and any other pertinent information. Once the specimen has been identified, another label should give the scientific name, sex (very important for ants), the name of the person who identified it, and the date it was identified. If it is re-identified as something else later on, then the old label should still be retained on the pin, but another superseding label added as well. Important specimens in old museum collections can easily acquire half a dozen data labels, and each adds to the information about the specimen and its provenance.

No museum curator has ever complained that a specimen has too many data labels underneath it or has too much information, and no information is too abstruse. It is a delight to browse museum collections and discover that a worker specimen of the subtropical *Polyrhachis lacteipennis* was plucked from a large ornamental date palm brought over from the United Arab Emirates for a show garden at the Chelsea Flower Show in June 2001 (now in the Royal Horticultural Society scientific collection), or that a damaged winged queen of *Myrmica scabrinodis* was hooked out of a cup of tea on Brighton Pier (now in the Booth Museum). My own personal contribution to the annals of data-label epistolography appears under a worker pharaoh ant: 'On fax machine, Fulham Road shop, 4.vii.1989'. Future myrmecologists will just have to check their history books to find out what a fax machine was.

If there were ever a reason to give too much seemingly obvious information on a data label, then it is the well-known fact that despite their seeming fixed nature in the world, places change, and so too do their names. When White (1895) illustrated the dulotic behaviour of *Formica sanguinea* on Shirley Common, Surrey, he and his readers knew exactly where he meant. But there is now no longer a place called Shirley in Surrey. There is one, however, in the London Borough of Croydon. Are these the same? They are, because, thankfully, British naturalists have long used a geographic identifier to bridge the gap of years – the vice-county (see box).

Storing ant specimens

Whatever the personal curation choices might be, it is important to house the reference collection in an airtight box so that museum/carpet beetles (*Anthrenus* species) and other pests cannot get in and destroy the preserved specimens. Every museum has horror stories of valuable insect specimens being destroyed because the chitin-eating larvae of these tiny insects (related to larder beetles) were quietly devouring away in the darkness; during the many months or years when the collection was not examined regularly, the collection was reduced to piles of dust surrounding naked pins and empty rectangles of card. Airtight boxes will also help keep out damp, which can encourage damaging moulds. Purpose-built wooden store-boxes can easily cost £50 each new, but they are still cheaper than museum-quality insect cabinets, and second-hand boxes are often available, particularly from museums shedding their excess stock of non-standard sizes. Historically they were

How to study ants

Where do you really live? A personal aside on vice-counties

In Britain the unique vice-county system of biological recording is at once both quaintly archaic and highly practical. It was devised by botanist H. C. Watson in 1852 and was based on the political county boundaries of the day, but sometimes subdivided or adjusted to give geographical zones all of approximately equal area. Thus Surrey, where I now live in East Dulwich, is vice-county 17, but Sussex is divided into East (14) where I grew up, and West (13), which I sometimes visited.

There have been significant boundary changes since that time – some to fit changing parliamentary constituencies (including gerrymandering perhaps) and later rearrangements based on local government administration and local authority funding zones (the abolition of Middlesex, for example, as it was swallowed up by Greater London in 1965). But using these original 1852 vice-counties as recording units lends a unity and geographic continuity. The *Ants of Surrey* (Pontin 2005) records the 30 species known from vice-county 17, which still includes the south-west quadrant of what is now Greater London, right up to the Thames near Greenwich. It also still retains the Croydon suburb of Shirley, and White's record of *Formica sanguinea* is included in the book. This modern Surrey list can be compared to the 24 species in the *Victoria County History* volume for Surrey (Saunders 1902).

Matching the biological vice-county to existing modern geography can, however, be a stretch of the ingenuity. Even 50 years after Watson adopted them, boundary movements had altered those 1852 county edges, so anyone needing to know the precise details of historical localities often had to resort to examining old maps. In 1969 the Ray Society published a national map showing the vice-county margins (Dandy 1969), but this was at such a small scale that it was still often difficult to interpret for anyone living on the edge. And very often the edges were convoluted beyond belief, following erratic field margins, tiny meandering streams, and blank open hillsides which seemed to have been divvied up between neighbouring hamlets arguing over grazing rights since the Dark Ages. A series of 1-inch Ordnance Survey maps annotated by Dandy in 1947 is held in the Natural History Museum, and was available to be consulted in person, but this was all a bit of a faff really. They have recently been digitised, and websites such as herbariaunited.org now offer interactive maps allowing vice-county detectives to make decisions about where they really stand.

There is still plenty of room for confusion, though. In 1996, when living in Nunhead, in south-east London, I noticed an old cast-iron post in my neighbour's garden. I could clearly make out part of the text 'WISHAM' up the side of it and realised that it was marking the old parish boundary between Lewisham (historically in Kent) and Camberwell (originally in Surrey). A bit of sleuthing soon showed that the parish and county boundaries did not quite coincide hereabouts, but instead diverged at that marker post, leaving most of what should have been a Lewisham (Kentish) field actually just inside Surrey (Jones 1997). My garden straddled this conundrum. It turns out that the vice-county boundary between Surrey (VC17) and West Kent (VC16) still ran right through the back 2m of my garden – it was quite a challenge correctly labelling specimens I found there.

cork-lined, then papered over, to take the pinned insects, but modern boxes use Plastazote sheets – which being white do not need to be laboriously papered over to make them look nice. I've been collecting ants for 45 years and my several hundred specimens still fit easily into two slightly scuffed second-hand corked store-boxes. The first specimen I bothered to take a second look at by collecting it to check under the microscope (except I didn't have one at the time – I was only 17, so I used a hand lens) was a worker of *Lasius fuliginosus* from Isfield, East Sussex, which I found on 4 October 1975, visiting Ivy *Hedera helix* flowers growing up the tree stump in which it was nesting. Although I knew the black pavement ant *Lasius niger*, I thought this looked different. And it was.

Once named using identification keys (see Appendix), the ant specimens can be displayed in some sort of taxonomic order, using the latest checklist names. This helps when later comparing new specimens to previously identified examples, or when double-checking if some new revision is published or species split announced. Inevitably there are difficulties with some specimens, especially where comparative measurements are required. The shape of the antennal scapes of *Myrmica* can be very confusing, and there is still debate about the validity of some species. When the difference between one species and the next is down to a few hairs on the back of the head, or the relative proportions of the eyes, it is sometimes difficult to be sure what you are looking at until you have a selection of specimens collected from several localities, often over many years, when finally the comparative differences can be demonstrated to your own satisfaction.

Keeping captive ant colonies

Ants have also lent themselves to the study of their captive colonies. For many years the invertebrate hall of London's Natural History Museum had a small colony of Trinidadian leafcutters on display, on two stone islands surrounded by water in a large glass display case. Branches of leaves dumped into one end of the glass case were dismembered and chopped up and transported along a rope bridge to the main nest on the other island. This might be a bit ambitious for the home laboratory, but 'ant-farm' or 'ant-town' formicaria are popular display items in homes and schools, and many different designs can be bought or built. All that is required is a container with soil, leaf litter or whatever, some means of preventing ant

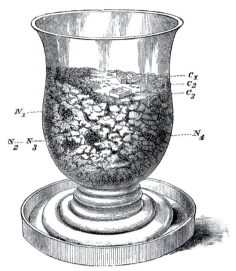

escapes, and a ready supply of food which can be topped up regularly. Standard observation formats are based around two sheets of glass separated by wooden or plastic spacers. The void between the glass is filled with sand or soil and the burrowings of the ants can then be viewed as they excavate tunnels and chambers. In 1996 Japanese artist Yukinori Yanagi erected a display of 49 interconnected formicarium chambers, narrow sand-filled plastic boxes, called *Pacific*. Each segment used coloured sands to recreate the flag of a nation bordering the Pacific Ocean; the ants gradually deconstructed and smeared the designs as they tunnelled through the sand. Open nests can be surrounded by a moat of water (like that of Swammerdam, see page 218), into which the ants are loath to venture. Alternatively a barrier of petroleum jelly or heavy mineral oil will stop the ants escaping, or they can be completely enclosed in a sealed container.

One of the simplest formicaria to construct uses just two glass jars. A large outer jar, like those used to preserve fruit or store foodstuffs, is part-filled with a sand/soil mix, and a slightly smaller jar is then inserted. The gap between the jars, ideally about 10–20mm, can then be infilled, and this forms the narrow soil-filled space in which the ants can burrow. Alternating thin layers of sand and soil will soon

ABOVE: Formicaria: left: a rather elaborate construction from Rennie's *Insect Architecture* (1845) and, right: one in the form of a simple glass vase from *Ants and Their Ways* (White 1895). White's labels refer to the several nest chambers containing pupae (N) and 'card trays used as cemeteries' (C). Having used these small trays to place food into the colony, he later observed up to 180 dead ants in one of them. Today we might unceremoniously describe them as middens.

demonstrate how the particles are being excavated and moved about. The smaller jar can also be used as a forage arena, to supply morsels of food, as long as the ants have a twig or two to climb up and out into the inter-jar nest zone. Topped with a piece of muslin or old net curtain held in place by an elastic band, the ants have no escape.

For the very smallest ant species, plaster of Paris poured into a broad dish can be carved as it nears setting point, to shape out small shallow galleries with interconnecting tunnels and a foraging arena. When fully hardened it can be covered with a tight-fitting sheet of glass and the ants released into it. Wood-nesting species can be housed in glass vivarium-type tanks, either with the nest log raised on a stone surrounded by water, like the Natural History Museum's leafcutters, or with a close-fitting lid.

An observation colony needs a queen and a cohort of her workers (and preferably some of the original nest material) to last any length of time – many years in some laboratory nests. A newly mated founding queen from a flying ant day can start a colony on her own, but it will take her several weeks in the small claustral nest before she rears her first brood of workers and any real digging or foraging activity can be seen. Short-term observations of burrowing can be made with just a collection of 20–150 workers; they will dig tunnels and forage the food placed out for them, but without a queen egg-layer to replenish them, the pseudo-colony will gradually wane and eventually fail. Colonies should not be crowded either; a general rule of thumb is that a nest should occupy only about 1% of the available space in the formicarium, giving the ants at least some feeling of naturalness in the amount of space available for foraging or territorial exploration.

Small colonies kept in house or school are likely to be much warmer than is natural. This does mean that activity can probably be observed year-long, but it also means that nest humidity may need adjustment. Small balls of cotton wool, scrunched tissue or wood shavings, kept moist but not wet, can be provided in part of the foraging arena and the ants will collect the moisture as they need it. And since ants are mostly subterranean, any observation nest should be kept covered up most of the time, keeping light out, and should only be uncovered when being directly observed or demonstrated to the schoolchildren. Thanks to the experimental diligence of various myrmecologists, an artificial diet can now be cooked in the kitchen (see box opposite), without resort to the parallel captive breeding of

Cooking up ant treats

Pet ants need pet food. This brew, first described by Bhatkar &
Whitcomb (1970), seems one of the most popular. They tested it on
several species of ants and almost all of them successfully matured
a new generation of sexuals – queens and males. Some species can
survive shorter or longer periods simply on sugar- or honey-water
supplemented with small bits of insect such as mealworms, fruit flies
or crickets.

Ingredients

- 1 egg
- 62ml honey (seems a bit too precise, but that's what they claim)
- 1 vitamin/mineral pill ground to a powder (any generic
  supermarket multivitamin pill/capsule)
- 5g agar powder (vegan gelatine alternative, in the home-baking
  aisle of the supermarket)
- 500ml water

Method

1. Dissolve the agar in 250ml of boiling water and leave to cool.
2. Blitz remaining 250ml of water and the rest of the ingredients in a
   food processor.
3. Combine everything together and poor approximately 5mm deep
   into several shallow vessels, such as petri dishes, to store in the
   fridge for a week or two.
4. Cut into small pieces about 1–2cm square for feeding.

aphid colonies for their honeydew, or having to drop in a constant
supply of live insect prey.

On a separate note, it is also easy to rear captive antlions, and
to watch their interactions with ants. All that's required is a small
container to take some of the loose dry sand in which the antlion is
living. Within half an hour of dropping the antlion into the sand arena,
it will have dug its conical pitfall trap. It must be fed several times a
day – it will, however, only accept live food. I experimented with
various ant sizes when visiting relatives in Florida many years ago.
At first my hosts were awed by the bizarre shape of the creature and
its excavation skills, but they soon became squeamishly revolted as I
tossed in live ants for it to devour mercilessly. Everyone was relieved
when digging stopped and the pit subsided into a vague dimple as

the insect pupated in the soil. They also showed less than correct enthusiasm when the rather drab winged adult emerged a few weeks later, after I had returned home, and they were faced with a strange fluttering critter crawling up the wall.

Spreading the word

Entomology can be a solitary pursuit, but even the most commonplace observations are interesting to discuss with other like-minded souls. It is easy for the newcomer to imagine that they will have nothing to contribute, because all the experts must have seen it all before. But any expert will soon tell you how little they really know, and will be fascinated to discover the minutiae of any personal observations. Until recently these observations were mostly published in specialist entomological journals, and these continue to be an important source of biological and geographical information. I just undertook a more or less random search through the *Entomologist's Monthly Magazine* from the 1860s and 1870s on the shelves of my office, and it presented the discovery of *Myrmica lobicornis* new to the Durham and Northumberland area (Bold 1866), a report of unseasonal production of winged garden ants *Lasius niger* ('*Formica nigra*') in the heated cactus house of the Cambridge Botanical Gardens ('many male and female specimens were struggling in the webs of sundry gaunt, hungry-looking spiders'; Eaton 1869), and the discovery of the blind mouse-nest beetle *Leptinus testaceus* in a nest of *L. fuliginosus* in Tilgate Forest (Champion 1870). We only know of these snippets, still fascinating after a century and a half, because the observers took the trouble to write up short notes and send them in for publication. It is such observations that are the bread and butter of field entomology.

When I noticed that the black ants in my front garden were attracted to damaged leaves of *Ligularia dentata* (Asteraceae), I first thought nothing of it. But then I remembered the received wisdom that this species subsisted only on the honeydew offerings of the aphids they milked. On hands and knees, and through the viewfinder of a camera, I was able to see that they were feeding directly on the crystallised sap exudations. Pleased with the photos I took, and wondering if this was an interesting new observation, I discussed it with colleagues at a meeting of the British Entomological and Natural History Society. Andrew Halstead, then entomologist with the Royal Horticultural Society garden at Wisley, commented that he had seen

a photograph in their archives of an unnamed brownish ant nibbling the leaves of a greenhouse plant, *Catharanthus roseus*, and had also seen *Lasius fuliginosus* chewing rose petals. Incorporating these snippets, I published a short note in a scientific journal (Jones 1994). By way of experiment I today (18 February 2020) sent out a request on Twitter asking for any similar information. Within minutes I had several responses from Surrey, Norfolk, Cambridgeshire and Texas. They commented on the similarity of this behaviour and ants drinking at extrafloral nectaries, and produced a link to the paper by Banks *et al.* (1991) describing the infamous fire ants *Solenopsis invicta* attacking citrus trees for the dribbling sap, often girdling the trees, which then died. I still do not know if ants feeding directly on dribbling plant sap is a widespread though overlooked behaviour, but when a major monograph on insect/plant interactions was published (Jolivet 1998), my humble two-page article was quoted in the introduction. I was pretty smug when I found that out.

Today there are plenty of specialist entomological publications always on the lookout for short notes of strange or unusual observations. There are also plenty of social media platforms where novice and expert communicate as equals and where easy exchange of information and enthusiasm takes place at the touch of a phone screen. The Bees, Wasps and Ants Recording Society (bwars.com) is a useful hub; it publishes a newsletter, has a friendly

Three *Lasius niger* workers feeding on crystallised sap where it was seeping from the damaged edge of a *Ligularia* leaf. This unusual behaviour has seldom been reported.

Portrait of a queen – a flight-ready *Lasius niger* female. There is still much to find out about when ants fly, how and where new nests are founded and how long queens live.

Facebook presence for people to post pictures for identification and comment, and its web pages offer useful identification guidance, ecological and distributional information. Meanwhile antwiki.org and antweb.org are veritable treasure troves of international information and expertise. As suggested earlier, recording even common and widespread species is still important, as the more records come in, the greater the detail and robustness of the distribution data. BWARS collates records sent in through its various recording schemes, but one-off records can be sent by the iRecord app from computers, tablets and smartphones. A phone call, text message or tweet can also send the information to other interested parties, and I'm pretty sure the traditional method of writing something on a postcard and slipping it into a pillar box will also still work too. We cannot always know if the casual records we make today are new or ground-breaking, but if we do nothing with them, nothing will happen.

Despite the sometimes slightly eccentric nature of the introvert entomologist (and I'm including myself in here too), the rule for the 21st-century naturalist ought to be 'Don't make a stranger of yourself'. Tell people what you find, what you see, and why you think it is interesting. I promise you that you'll receive a warm welcome and enthusiastic response.

Ants, it turns out, are certainly very small, but the study of ants is a vast subject encapsulating almost everything that humans know about the way life on this planet works. Their prodigious numbers, their frenzied activity, and their complex social coordination behaviours lift ants from their lowly earth-bound existence and take them to exalted scientific heights. There is always something new to be learned. And ants are everywhere, available and amenable to study by expert and beginner alike. This book is but a drop in the myrmecological ocean; nevertheless I hope it will give a flavour of the wonders to be seen, the insights to be discovered, and the fascination to be gained. In conclusion I offer an updated version of King Solomon's proverbial pronouncement:

Go to the ant, don't be a sluggard; consider her ways, and come away wiser.

Ants in waiting – *Myrmica rubra* larvae and pupae showing the 'naked' chrysalis, lacking a silken cocoon. Brood times, emergence behaviours and size variation in workers are all potential study areas.

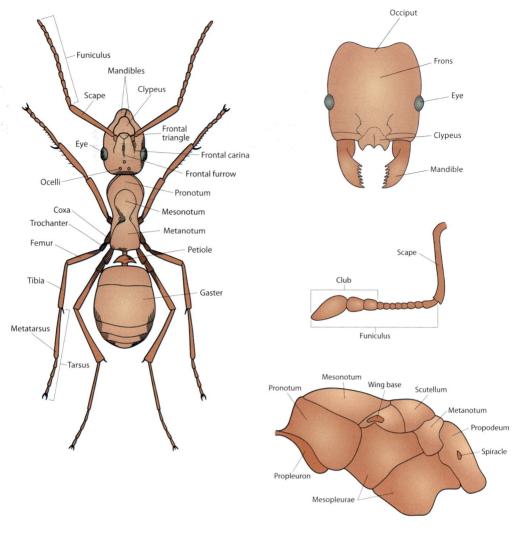
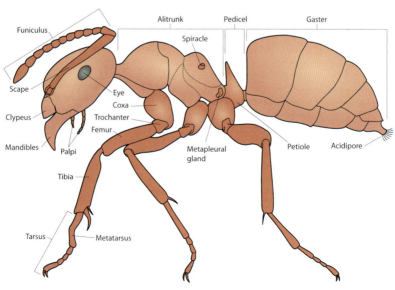

Identification key

E xcellent, though sometimes highly technical, identification keys are available to any readers wishing to take their myrmecological interest further. I'd recommend Bolton & Collingwood (1975), Collingwood (1979), Skinner & Allen (1996) and Lebas *et al.* (2019); the first two are currently available as PDFs freely downloadable from the internet. Looking beyond the British Isles, five modern treatments are also available for the ants of Poland (Czechowski *et al.* 2002, 2012), central and northern Europe (Seifert 2007; in German, translated into English 2018), France (Blatrix *et al.* 2013; in French) and Wallonia (Wegnez *et al.* 2012, also in French). Other useful works include those by Sweeney (1950), Bernard (1968), the important work on *Formica* by Yarrow (1954), a key to *Lasius* by Blacker & Collingwood (2002) and one to *Myrmica* workers by Bolton (2005). A useful key to ant species found in Surrey is also given by Pontin (2005). The following identification key is built on the work of these previous authors and some of the thumbnail figures are redrawn, with permission, from these works.

Because of ants' complex caste systems, there is usually a require-ment for three separate identification keys – one for workers, one for queens and one for males. This can make for repetitive reading and often a wasteful duplication (or triplication) of diagrams. Therefore, what follows is a key to only the workers of British species – and a few others. I hope it suits. There is, inevitably, a need to look very closely at an ant specimen to secure a confident identification. Whether specimens are collected, killed, and mounted for detailed examination under the stereomicroscope (see Chapter 10) is a moral, ethical and scientific decision each reader must make according to their own circumstances. The key that follows is less complex than many, but still, perhaps, daunting for the beginner. My advice: have a go. The difference between a casual observer and an expert is a long one, but anyone and everyone can aspire to have a closer look, to gain deeper understanding, to see for themselves rather than rely on

OPPOSITE PAGE:
Diagrams of a typical ant. Clockwise from top left: body top view; head; geniculate antenna; mesosoma/alitrunk side view; body side view.

regurgitated information from distant and highfalutin monographs or coffee-table picture books. Even under a cheap hand lens, ants (indeed all insects) are revealed in a wondrous detail hidden from the naked eye. The following key is for known British and Irish species, plus a few others that have either become naturalised here, or are known from nearby Europe and which could turn up any day now, plus a few rather speculative species that might appear here in the next couple of decades.

Key to the castes of British and Irish ant species

After you have found an ant nest, it should be pretty obvious which are the worker ants – they're the ones rushing about all over the place doing the work. Occasionally, though, a lone ant turns up, and although these are usually workers it may be necessary to double-check.

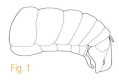

Fig. 1

1   Gaster with 6 visible segments if the pedicel is 1-segmented (petiole only), or 5 visible segments if the pedicel is 2-segmented (petiole plus postpetiole). Tip of gaster with complex genital capsule projecting, lacking sting or acidopore (Figure 1).................. **Males**

Males of most British ants are winged and short-lived. They emerge to take part in nuptial flights but make no effort in colony formation and soon die or are eaten by predators.

—   Gaster with 5 visible segments if the pedicel is 1-segmented, or 4 visible segments if the pedicel is 2-segmented. Tip of gaster without projecting genital capsule, terminating in a sting (Myrmicinae, Ponerinae), acidopore (Formicinae) or a transverse groove (Dolichoderinae) .... **Females 2**

2   Wings present (at least in young individuals) or if missing (in mature queens) then alitrunk flight sclerites discernible and often torn wing base remnants visible beneath the tegulae .... **Queens**

—   Wings absent, alitrunk sclerites fused together to give semblance of smooth carapace lacking tegulae or wing-stub remnants.. **Workers**

Key to the workers of British and Irish ant species

1   Pedicel with just a single segment – the petiole (Figures 2, 3, 4) **32**

—   Pedicel with 2 segments – petiole and postpetiole (Figure 5). Subfamily **Myrmicinae** **2**

Fig. 2

2   Postpetiole attached to the upper surface of the first gaster segment (Figure 6). Gaster heart-shaped when viewed from above – broadly rounded at the base but smartly narrowed to a sharp point at the apex ....Genus *Crematogaster*

      A tree-dwelling species sometimes imported with cut wood, cork or timber from southern Europe or North Africa.

Fig. 3

—   Postpetiole attached to the front face of the gaster. Gaster not heart-shaped **3**

Fig. 4

3   Jaws narrow, curved and pointed, lacking teeth along their inner edges (Figure 7), or very long and straight with a single tooth just before apex (Figure 8). Rare species............**4**

—   Jaws broad, triangular with several teeth along their inner edges (Figure 9) **6**

Fig. 5

4   Head long, heart-shaped, temples very strongly rounded (Figure 8). Jaws very long, straight and narrow, held nearly parallel, strong single narrow tooth just before apex...*Strumigenys perplexa*

      Tiny (1.5–2.0mm) trap-jaw ant able to open mandibles to 180 degrees. Recently discovered in Guernsey.

Fig. 6

—   Head shorter, rectangular or triangular (Figure 7). Jaws slim but not long and parallel **5**

5   Antennae with 12 segments (scape and 11 funicular segments). Head long, mandibles more prominent ....*Strongylognathus testaceus*

      A rare social parasite in nests of Tetramorium caespitum.

Fig. 7

—   Antennae with 10 or 11 segments. Head broad, mandibles less prominent ....*Tetramorium (Anergates) atratulum*

      A rare social parasite in nests of Tetramorium caespitum. *It has no worker caste, so small wingless queens should key out here.*

Fig. 8

6   Propodeum rounded, without projecting teeth or spines (Figure 10) **7**

—   Rear end of propodeum with a pair of teeth or spines (Figure 11) **9**

Fig. 9

Fig. 10

Fig. 11

7   Antenna with 10 segments (scape and 9 funicular), the last 2 forming
a small club (Figure 12) .... *Solenopsis fugax*

> *A rare and secretive southern species usually associated with nests
> of* Lasius *or* Formica *species.*

Fig. 12

—   Antenna with 12 segments (scape and 11 funicular) 8

8   Small, <2.5mm, and pale. Last 3 segments of funiculus forming
a small club .... *Monomorium pharaonis*

> *An introduced tramp species, only found in heated buildings.*

Fig. 13

—   Larger, 3.0–12.0mm, black. Antennae not clubbed .... *Messor capitatus*

> *A seed-harvesting ant, occurs in Brittany and captive colonies are kept by
> ant enthusiasts.*

Fig. 14

9   Antenna with 11 segments (scape and 10 funicular) 10

—   Antenna with 12 segments (scape and 11 funicular) 11

Fig. 15

10   Underside of postpetiole with distinct, sharp, forward-projecting spine
(Figure 13). Head and alitrunk smooth and shining above .... *Formicoxenus nitidulus*

> *A secretive species that forms small colonies inside the mound-nests of
> Formica rufa and F. lugubris.*

—   Underside of postpetiole without a spine (Figure 14). Head and alitrunk
sculptured with fine wrinkled grooves .... *Leptothorax acervorum*

> *Small colonies in tree stumps, under bark.*

Fig. 16

11   Underside of head with a low but sharp ridge along each side (Figure 15).
Front edge of clypeus with two prominent lobes or teeth, sometimes with a
third, smaller, tooth in the middle (Figure 16) .... *Myrmecina graminicola*

> *Small colonies under stones and logs.*

Fig. 17

—   Underside of head without ridge on each side. Front edge of clypeus without teeth 12

12   Side margins of clypeus with their hind edges raised into ridges in front of the
antennal sockets (Figure 17). Shoulders of pronotum prominent,
almost angulate (Figure 18) .... *Tetramorium caespitum*

> *Mostly coastal, southern or heathlands.*

Fig. 18

—   Sides of clypeus not raised into prominent ridges along hind edges. Shoulders of
pronotum more rounded 13

Fig. 19

13   Petiole long, more stalk-like where it joins the propodeum (Figure 19), and with a
blunt subconical node 14

—   Petiole shorter and stouter (Figure 20) 16

Fig. 20

14   In profile back of alitrunk appearing level and flat, mesonotum not strongly
arched (Figure 21). Propodeal spines well developed .... Genus *Stenamma* 15

Fig. 21

— In profile mesonotum appearing strongly arched (Figure 22). Propodeal spines
short and blunt...*Aphaenogaster subterranea*

Widespread in Europe, potential future colonist.

Fig. 22

15 Petiole slightly constricted near base, with small tubercle on each side (Figure 23).
Frontal triangle broader, more parallel-sided, U-shaped, about one-third distance
between antennal bases. Legs shorter and stouter.................................*Stenamma debile*

Widespread. Only recently separated from the following species.

Fig. 23

— Petiole more smoothly narrowed to base, lacking tubercles (Figure 24). Frontal triangle
narrower, slightly constricted in middle, less than one-third distance between antennal
bases. Legs longer and slimmer ....*Stenamma westwoodii*

Apparently rare. Only recently separated from the previous species.

Fig. 24

16 Workers of two different sizes, the larger majors with huge head and mandibles with
only three teeth, one at the base and two at the tip, separated by a long straight
cutting edge. Head and alitrunk shining ....*Pheidole megacephala*

Introduced species in buildings, hothouses etc.

Fig. 25

— Workers all the same size, none with only three teeth on mandibles. Head and alitrunk
usually wrinkled and sculptured..17

Fig. 26

17 Antenna with last three segments larger, nearly as long as the rest of the funiculus
(Figure 25). Antennal scapes without long curved hairs. Body hairs stout and blunt. Smaller
species, usually <3.5mm. ....Genus *Temnothorax* 18

— Antenna with last segments less obviously forming a club, not as long as the rest
of the funiculus (Figure 26). Antennal scapes with long curved hairs. Body hairs fine
and sharp. Larger species, usually >3.5mm. ....Genus *Myrmica* 21

Fig. 27

18 Antennae unicolorous reddish, apical antennal segments not darker than
remainder of funiculus...*Temnothorax nylanderi*

Small colonies in tree stumps, under bark.

Fig. 28

— Apical antennal segments darker than remainder of the funiculus 19

19 Propodeal spines short and straight, only about half as long as the distance
between them (Figure 27) ....*Temnothorax albipennis*

Rare, southern, coastal.

Fig. 29

— Propodeal spines longer and curved slightly inwards, about as long as the distance
between them (Figure 28) 20

Fig. 30

20 Top of head between eyes with sculpture consisting of longitudinal wrinkles
(Figure 29). Propodeal spines longer. Gaster pale or dark, but without a distinct
band across it (Figures 30, 31) ....*Temnothorax interruptus*

Rare, southern, coastal.

Fig. 31

— Top of head with sculpture showing many fine cross links between the longitudinal wrinkles, giving a more net-like pattern (Figure 32). Propodeal spines short (but not as short as *T. albipennis*). Gaster usually marked with a distinct dark band across the middle (Figure 33) .... *Temnothorax unifasciatus*

 Channel Isles.

Fig. 32

21 Underside of postpetiole with a large forwards-directed projection easily seen in side view (Figure 34) .... *Myrmica karavajevi*

 A rare social parasite in the nests of *M. sabuleti* and M. scabrinodis. *It has no worker caste, but small wingless queens should key out here.*

Fig. 33

— Underside of postpetiole without a large projection... 22

Fig. 34

22 Postpetiole very broad, much broader than petiole (Figure 35). Body covered with long hairs .... *Myrmica hirsuta*

 A very rare (Dorset and Hampshire) social parasite in nests of *M. sabuleti.* *Workers are extremely rare, but small wingless queens should key out here.*

Fig. 35

— Postpetiole similar width to petiole (Figure 36). Body not as hairy 23

23 Scape of antenna gently curved at base, this curve might be quite sudden or tight, but it is not sharp almost right-angled, nor is it armed with a node, tooth, flange or outgrowth on this corner (Figure 37).. 24

Fig. 36

— Scape of antenna abruptly angled at base, the angle being sharp, almost right-angled (Figure 38), with the corner of the bend sometimes armed with a node, tooth, flange or outgrowth (Figures 39, 40)... 26

Fig. 37

24 Scape more suddenly curved. Frontal triangle sculptured with several longitudinal wrinkles..*Myrmica sulcinodis*

 Moorland, heathland, mountains.

Fig. 38

— Scape more gently curved. Frontal triangle smooth 25

Fig. 39

25 Rear surface of the petiole more smoothly curved from its upper dome to its hind edge so that it appears sinuously shaped in profile (Figure 41). Propodeal spines shorter, less than the distance between their tips ....*Myrmica rubra*

 Widespread.

Fig. 40

— Rear surface of the petiole appearing notched in profile, where the flatter dome suddenly drops away behind, then equally suddenly flattens out again (Figure 42). Propodeal spines longer, greater than the distance between their tips... *Myrmica ruginodis*

 Widespread.

Fig. 41

26 Scape simply right-angled, without node, tooth, flange or outgrowth (Figure 38) 27

— Scape with node, tooth, flange or large outgrowth at the bend (Figures 39, 40)...............29

Fig. 42

27 Rear surface of the petiole smoothly curved from its upper dome to its hind edge so that it appears sinuously shaped in profile (Figure 43) ....*Myrmica specioides*

    Rare, southern, coastal.

Fig. 43

— Rear surface of the petiole appearing notched in profile, where the flatter dome suddenly drops away behind, then equally suddenly flattens out again (as in Figure 42) **28**

28 Clypeus with small median notch. Slightly more hairy – petiole with 10–20 standing hairs. Alitrunk with less rugose and almost straight longitudinal sculpture (Figure 44)...*Myrmica vandeli*

    Rare, variable, poorly understood species.

Fig. 44

— Clypeus smoothly rounded in front. Less hairy – petiole with less than 10 hairs. Alitrunk with coarser strongly reticulate sculpture (Figure 45).........................*Myrmica scabrinodis*

    Widespread.

Fig. 45

29 Scape stouter, with a large flattened outgrowth at the corner extending into a longitudinal keel running along the edge (Figure 39).. **30**

— Scape more slender, with a smaller, tooth-like ridge or flange across the corner angle..... **31**

30 Scape with very large outgrowth at basal corner .... *Myrmica lonae*

    Status uncertain but apparently scarce.

Fig. 46

— Scape with smaller outgrowth at basal corner .... *Myrmica sabuleti*

    Status uncertain but apparently widespread.

31 Ridge across the corner of the scape larger, a broad rounded flange. Frontal carinae distorted to accommodate this structure (Figure 46)...............................*Myrmica schencki*

    Local, southern.

Fig. 47

— Ridge at the corner of the scape narrower, pointed, more tooth-like. Frontal carinae not distorted (Figure 47).. *Myrmica lobicornis*

    Widespread.

Fig. 48

32 Gaster clearly constricted between first and second segments (Figure 48). Subfamily **Ponerinae** **33**

— Gaster not clearly constricted between first and second segments.................................... **35**

33 Underside of petiole with a small backward-pointing tooth or hook towards the rear on each side, and with a small translucent fenestral window spot towards the front (Figure 49).. **34**

Fig. 49

— Underside of petiole without a tooth or hook on each side and lacking any translucent spot (Figure 50)..*Hyperponera punctatissima*

    Rare, mostly in buildings.

Fig. 50

34 Colour dark, black or deep brownish red. Petiole taller and narrower (Figure 51), more nearly smoothly eliptical in cross-section..*Ponera coarctata*

   Local, southern.

— Colour reddish brown to pinkish yellow. Petiole shorter and thicker (Figure 52), approximately semicircular in cross-section ....*Ponera testacea*

   Rare, coastal, only recently discovered in the British Isles.

35 In side view, the tip of the gaster with a conical structure surrounded by short hairs – the acidopore (Figures 53, 54). Some standing hairs on top of alitrunk and head. Subfamily **Formicinae** **40**

— Gaster not tipped with an acidopore, ending with a transverse slit (Figures 55, 56). Alitrunk, and head behind the clypeus, lacking standing hairs. Subfamily **Dolichoderinae**. [Scarce species, or those only occurring in heated buildings. If in doubt, go to couplet 40] **36**

36 Petiole small and rather cylindrical, usually hidden beneath overhanging front edge of the gaster (Figure 57) **37**

— Petiole large, flattened and scale-like, usually easily visible from above (Figure 58)...*Linepithema humile*

   In heated buildings.

37 Body moderately shining black with strongly contrasting white tarsi. Pronotum and gaster with several long stout bristles arising from the short pubescence ....*Technomyrmex albipes*

   In heated buildings.

— Not black with strongly contrasting white tarsi, although may be dark brown with pale brown appendages. Pronotum and gaster only with short pubescence ....Genus *Tapinoma* **38**

38 Bicoloured, gaster pale yellow, sometimes almost white, head and alitrunk dark brown or blackish. Legs and antennae pale ....*Tapinoma melanocephalum*

   In heated buildings.

— Unicolorous black or very dark brown.. **39**

39 Front edge of clypeus with a deep notch (Figure 59). Vaguely larger ......*Tapinoma erraticum*

   Rare, southern, coastal.

— Front edge of clypeus with a shallow notch (Figure 60). Vaguely smaller.. *Tapinoma subboreale*

   Rare, southern, coastal, only recently separated from previous species.

40 Antennae with 11 segments (scape plus 10 funicular segments)............*Plagiolepis taurica*

   Channel Islands.

Fig. 51

Fig. 52

Fig. 53

Fig.54

Fig. 55

Fig. 56

Fig. 57

Fig. 58

Fig. 59

Fig. 60

— Antennae with 12 segments (scape plus 11 funicular segments)....................................... 41

41 Mandibles slim and curved like sabres (Figure 61)............................. *Polyergus rufescens*
   Potential colonist from mainland Europe.

Fig. 61

— Mandibles broad triangular, toothed along inner edges 42

42 Antennae inserted into articulations slightly removed behind the hind edge
   of the clypeus (Figure 62)... Genus *Camponotus*
   Occasional import with timber, potential colonist from mainland Europe.

Fig. 62

— Antennal articulations very close to or touching the hind edge of the clypeus.................. 43

43 Eyes at or in front of middle of sides of head. Petiole sloping forwards and overhung
   by front edge of gaster. Rare imports, only in heated buildings... 44

— Eyes set behind middle of sides of head. Petiole not sloping forwards, scale-like, not
   overhung by front edge of gaster. Native outdoor species 45

Fig. 63

44 Antennal scapes very long, reaching backwards as far as the propodeum
   (Figure 63). Hairs on body (especially those on legs) long and sparse.
   Scapes without hairs..*Paratrechina longicornis*
   Occasional import, only able to survive in heated buildings.

Fig. 64

— Antennal scapes shorter, reaching backwards as far as the mesonotum (Figure 64).
   Hairs on body short and dense, including on scape .... *Nylanderia vividula*
   Occasional import, only able to survive in heated buildings.

Fig. 65

45 Opening of the propodeal spiracle circular or broad oval; in side view it is situated high
   on the junction of side and rear edge of the propodeum (Figure 65). Hind tibiae with no
   row of bristles on the underside, but there are bristles at the tips, and they are covered
   all over with short soft hairs. Funicular segments longer, segments 2–5 shorter than
   segments 6–10. Generally smaller, 2.2–6.0mm. .... Genus *Lasius* 46

Fig. 66

— Opening of the propodeal spiracle narrow, often slit-like; in side view it is situated in
   the middle of the side of the propodeum (Figure 66). Hind tibiae with a double row of
   bristles on the underside. Funiculus segments longer, segments 2–5 longer than
   segments 6–10. Generally larger, 4.0–9.5mm.....................................Genus *Formica* 59

46 Shining black. Head large, heart-shaped with rear margin deeply indented
   (Figure 67).. *Lasius fuliginosus*
   Widespread, nests in trees.

Fig. 67

— Not shining, colour dull black, yellow, brown, greyish, sometimes bicoloured.................. 47

47 Yellow or pale brownish. Eyes smaller, less than one-sixth of head width, and less
   than length of last antennal segment (Figure 68)... 48

Fig. 68

— Grey, black or brown, although alitrunk sometimes lighter than head and gaster. Eyes larger, more than one-fifth width of head; about equal to length of last antennal segment (Figure 69) **54**

Fig. 69

48 Antennal scapes (and front tibiae) with longer erect or suberect hairs standing out from the shorter pubescence (Figure 70) **49**

— Antennal scapes (and front tibiae) only with shorter pubescence (Figure 71) **52**

Fig. 70

49 Strongly bicoloured; alitrunk bright red and contrasting with the much darker head and gaster .... *Lasius emarginatus*

   Channel Islands, only recently on mainland Britain.

Fig. 71

— Head, alitrunk and gaster dull pale brown or yellow **50**

50 Antennal scapes with the longer hairs sloping. Erect hairs on gaster shorter. Hind tibiae with only a few erect hairs ....*Lasius sabularum*

   Rare.

— Antennal scapes with the longer hairs upright. Erect hairs on gaster longer. Hind tibiae with numerous erect hairs **51**

51 Antennal scapes and tibiae slightly flattened in cross-section; mid-point about twice as wide as thick. Petiole slightly flattened across top, subrectangular. Funiculus segments longer ....*Lasius meridionalis*

   Rare.

— Antennal scapes nearly circular in cross-section, at most one-and-a-half times as wide as thick. Petiole more rounded at sides and top. Funiculus segments shorter and more close-fitting together .... *Lasius umbratus*

   Widespread.

Fig. 72

52 Body hairs on gaster long (Figure 72). Underside of head lacking erect hairs **53**

— Body hairs on gaster short (Figure 73). Underside of head with erect hairs ....*Lasius mixtus*

   Uncommon.

Fig. 73

53 Eyes larger... *Lasius flavus*

   Very common.

— Eyes smaller .... *Lasius myops*

   Status uncertain in the British Isles.

54 Antennal scapes and tibiae with numerous outstanding erect hairs **55**

— Antennal scapes and tibiae with only short pubescence...**56**

55 Body hairs denser, particularly noticeable on the clypeus (Figure 74), which is conspicuously arched ....*Lasius niger*

    Very common and widespread.

— Body hairs less dense, particularly noticeable on the flatter clypeus (Figure 75)...*Lasius platythorax*

    Widespread, only recently separated from L. niger.

Fig. 74

Fig. 75

56 Head, and especially alitrunk, reddish, paler than the dark brown gaster. Head broad, nearly as wide as long. Frontal triangle well defined ....*Lasius brunneus*

    Local, nests in trees.

— Head, alitrunk and gaster more or less uniformly dark. Head narrower; clearly longer than broad. Frontal triangle indistinct **57**

57 Each mandible with 7 teeth, basal end with 2 large teeth (Figure 76) ....*Lasius neglectus*

    Invasive species, recently discovered in Britain, in gardens.

— Each mandible with 8 teeth, basal end with 3 large teeth (Figure 77) **58**

Fig. 76

Fig. 77

58 Hind tibiae with 2 or 3 slightly erect hairs. Rear slope of the propodeum with short hairs...*Lasius psammophilus*

    Local, only recently described.

— Hind tibiae only with short pubescence. Rear slope of propodeum without short hairs ....*Lasius alienus*

    Local, mostly coastal.

59 Front edge of clypeus with a distinct notch (Figure 78) .... *Formica sanguinea*

    Local, heaths, moors.

— Front edge of clypeus smoothly rounded **60**

Fig. 78

60 Rear of head deeply concave (Figure 79)..*Formica exsecta*

    Local, heaths, moors.

— Rear of head more or less straight, vaguely convex or very slightly concave....................**61**

Fig. 79

61 Alitrunk uniformly dark, brown or black, without paler reddish patches............................**62**

— Alitrunk with some reddish patches; rarely entirely red **64**

62 Shining black, sometimes vaguely brownish. Underside of head with two long hairs, though these are easily broken off... *Formica picea*

    Rare, lowland heaths and bogs.

— Not strongly shining; black or brownish. Underside of head without pair of long hairs.......**63**

63 Pronotum with numerous short erect hairs. Under surface of middle and
hind femora with numerous long hairs .... *Formica lemani*
  Local, northern, heaths, moors.

— Pronotum with at most only a few short erect hairs. Under surface of middle
and hind femora lacking long hairs .... *Formica fusca*
  Widespread in the south.

64 Frontal triangle dull, not shining **65**

— Frontal triangle mostly shining, unsculptured, or with a few fine punctures forming
a vague crescent-shaped mark across it **66**

65 Pronotum with numerous erect hairs...*Formica rufibarbis*
  Very rare, Surrey and Scilly.

— Pronotum with only four or fewer erect hairs .... *Formica cunicularia*
  Widespread.

66 Area of head between the frontal carinae matt under the microscope; almost
lacking any sculpture...*Formica pratensis*
  Channel Islands.

— Area of head between frontal carinae shining, although sculptured with fine punctures.... **67**

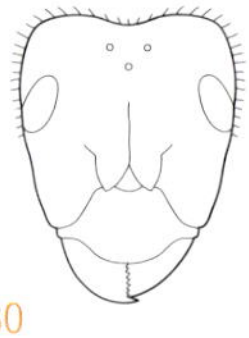

67 Eyes with minute hairs arising from between the facets. Hind margin of head with
outstanding erect hairs (Figures 80, 81)...**68**

Fig. 80

— Eyes without hairs. Hind margin of head without hairs (Figure 82)...................*Formica rufa*
  Widespread.

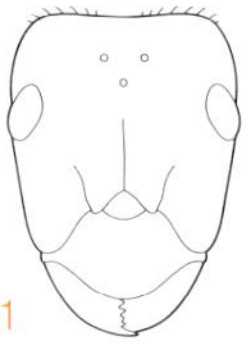

68 Hairs on hind margin of head very distinct and extending down the sides to the
eyes (Figure 80). Mesopleuron with numerous long hairs across it...........*Formica lugubris*
  Local, Scotland and northern England.

Fig. 81

— Hairs on hind margin of head less distinct, not extending beyond the rear corners
of the head and not reaching down to the level of the eyes (Figure 81).
Mesopleuron with long hairs only in lower front corner..........................*Formica aquilonia*
  Local, Scotland.

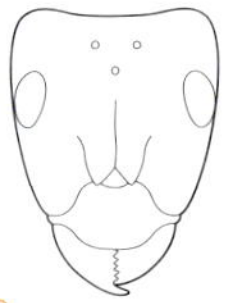

Fig. 82

OPPOSITE PAGE:
During the writing of this book the author slightly obsessed about ants, and took every opportunity to indoctrinate his family. Even at the breakfast table. So, pancake ants: **1** tiny ants, **2** generic wood ant, **3** trap-jaw ant *Odontomachus*, **4** bullet ant *Paraponera clavata*, **5** bullhead ant *Zacryptocerus*, **6** honeypot replete, **7** leafcutter, **8** ant woodlouse *Platyarthrus hoffmannseggii*, **9** ant beetle *Claviger testaceus*, **10** velvet (not an) ant *Mutilla europaea*.

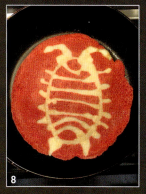

Glossary

acidopore, opening at the end of the ant gaster through which formic acid is squirted by formicine species.

alate, winged ant – male or female; a queen becomes dealate when she sheds her wings.

alitrunk, middle portion of ant's body comprising thorax and fused first segment of the abdomen (the propodeum). Also called mesosoma.

allometry, the growth of different body parts (usually in the pupa/chrysalis stage) at different rates, producing ants with different body proportions in the same species.

antennomere, single segment of an antenna.

Beltian body, nutrient-rich swelling on plant providing food for ants.

carina, raised keel-like ridge.

caste, one of a variety of distinctively different body forms within an ant species — usually represented by male, queen or worker, but also by workers of different body sizes and forms.

claustral, nest founding by single impregnated non-foraging queen confining herself in a hole.

clypeus, protective plate on the lower face over the jaw articulations.

cryptic species, species that is thought to be distinct but cannot readily be distinguished morphologically from another; or multiple possible species presently classified under a single scientific name.

diapause, period of suspended larval development, usually over winter.

dimorphic, having two types of body proportion among the workers, usually large majors and smaller minors.

diploid, having the full chromosome complement of pairs, typically resulting in female ant.

dulosis, capturing larvae and pupae from another ant species and rearing them to act as captive workers; also called 'slave-making'.

ergatoid, permanently wingless reproductive queen.

eusocial, social organisation of the colony that involves a numerous non-reproductive and ultimately disposable female worker caste.

formic acid, the simplest carboxylic acid, with formula HCOOH; smells slightly of vinegar.

frons, front of the head between the eyes.

funiculus, long flexible multi-segmented part of an ant's antenna.

gamergate, worker that is able to mate and lay viable eggs.

gaster, final bulbous segments of an ant's abdomen.

geniculate, elbowed, typified in ant antennae.

gongylidia, nutrient-rich swellings in strands of the fungus cultivated by leafcutter ants.

haemolymph, insect equivalent of blood, mostly carrying nutrients, but not involved with gas exchange.

hamulus, series of hooks on leading edge of hind wing attaching it to front wing.

haplodiploidy, sex determination through chromosome number; see diploid and haploid.

haploid, having half the full chromosome complement, one from each pair, typically resulting in male ant.

honeydew, liquid aphid excrement, more or less unchanged from plant sap; tastes like honey, looks like dew.

inquilinism, permanent social parasitism where one ant species moves in with another and lays its eggs for the host ants to rear; only males and queens are produced, it does not have its own worker caste.

instar, one of several discrete larval growth stages separated by skin moults, usually four in ants.

mandible, jaw.

mesonotum, second upper plate of the thorax.

mesosoma, middle portion of ant's body, comprising thorax and fused first segment of the abdomen; also called alitrunk, especially by myrmecologists.

metamerism, ancient evolutionary process that gave rise to multi-segmented bodies in animals.

metapleural gland, secretory gland on propodeum, unique to ants.

monogyny, having just one reproductive queen in the nest.

monomorphic, having just a single body-proportion type among the workers – most British ant species.

myrmecochory, provision of a part of a seed to be eaten by ants, encouraging them to collect and move the seed to the nest, thereby aiding distribution and avoiding other major seed-eaters.

myrmecology, study of ants.

myrmecophile, animal, usually invertebrate, that especially lives in ant nests or in close association with ants.

nanitic, smaller-than-average worker, usually the first cohort produced by a new queen.

nuptial flight, massed mating flight of winged male and female ants.

occiput, rear portion of the head, behind and between the eyes.

ocellus (plural **ocelli**), one of usually three simple light-sensitive domes on the top of the head.

ommatidium (plural **ommatidia**), single facet (and the light-sensitive column beneath) of an insect's compound eye.

pedicel, narrow one- or two-segmented hinge between propodeum and gaster, comprising petiole or petiole plus postpetiole.

petiole, second abdominal segment, reduced to small node hinge in ants.

pheromone, chemical scent released by one individual that affects the behaviour or physiology of another individual.

physogastry, grossly swollen abdomen of queen ant caused by massed eggs developing inside.

plerergate, honeypot ant, immobile worker whose sole purpose is to store liquid food in her extremely distended abdomen.

polydomy, several interconnected nests forming one ant colony.

polygyny, having several reproductive queens in the nest.

polymorphic, having at least three body-proportion types among the worker ants – for example major soldiers, media foragers and minor brood-tenders.

postpetiole, in Myrmecinae, the third abdominal segment, reduced to a small round node.

pronotum, first upper plate of the thorax, behind the head.

propodeum, first segment of the ant abdomen, fused to the thorax to make the combined body segment the mesosoma or alitrunk.

rugose, ridged, wrinkled or roughly sculptured.

scape, long thin first segment of an ant's antenna.

spermatheca, sperm-storage organ in the abdomen of fertile female (queen) ant.

staphyla, cluster of gongylidia, the nutrient-rich swellings in strands of the fungus cultivated by leafcutter ants.

sternite, lower body plate under each segment of the abdomen.

supercolony, large (huge) area of intermingling colonies that do not recognise each other as different because they all arose from a very shallow gene pool.

symphile, lodger in ant nests, fully accepted by the host ants and usually fed by them.

synechthran, opportunistic scavenger/predator in ant nests, treated with hostility by the host ants.

synoekete, scavenger/predator in ant nests, generally ignored by the host ants.

tagmosis, ancient evolutionary process where multiple body segments combined to form the head/thorax/abdomen form of extant insects.

tegula (pleural **tegulae**), small rounded epaulette-like plate covering the base of the wing in alate males and queens.

tergite, upper body plate of each segment of the abdomen.

thermophilic, warmth-loving, like most ants (and probably myrmecologists).

trichome, hair or long narrow scale sprouting from secretory pore.

trophallaxis, mouth-to-mouth exchange of liquid food between worker sisters, queen and brood in the nest.

trophic eggs, non-fertile eggs, usually lacking DNA, produced by a cloistered queen as nutritional packets to feed her first cohort of grubs.

trophobiont, aphid, mealybug or other insect 'farmed' by ants in or outside the nest.

trophobiotic organ, basket of bristles around an aphid's excretory vent to hold a droplet of honeydew ready for an ant to drink.

xenobiosis, one species of ant making its own independent nests inside the colony of another species.

References

Adams, E. S. 2016. Territoriality in ants (Hymenoptera: Formicidae): a review. *Myrmecological News* 23: 101–118.

Agrain, F. A. , Buffington, M. L., Chaboo, C. S., Chamorro, M. L., and Schöller, M. 2015. Leaf beetles are ant-nest beetles: the curious life of the juvenile stages of case-bearers (Coleoptera, Chrysomelidae, Cryptocephalinae). *Zookeys* 547: 133–164.

Alba-Lynn, C., and Henk, S. 2010. Potential for ants and vertebrate predators to shape seed-dispersal dynamics of the invasive thistles *Cirsium arvense* and *Carduus nutans* in their introduced range (North America). *Plant Ecology* 210: 291–301.

Alexander, K. N. A. 2006. *Lasius brunneus* (Latreille) (Hymenoptera: Formicidae) in Cambridgeshire. *British Journal of Entomology and Natural History* 19: 156.

Alexander, K. N. A. 2007. *Lasius flavus* (Fabr.) (Hymenoptera: Formicidae) nesting in heartwood decay in old orchard trees. *British Journal of Entomology and Natural History* 20: 35–36.

Alexander, K. N. A., and Taylor, A. 1998. The Severn Vale, a national stronghold for *Lasius brunneus* (Latreille) (Hymenoptera: Formicidae). *British Journal of Entomology and Natural History* 10: 217–219.

Allen, G. W. 2009. *Bees, Wasps and Ants of Kent*. Maidstone: Kent Field Club.

Allies, A. B. 1984. The cuckoo ant, *Leptothorax kutteri*: the behaviour and reproductive strategy of a workerless parasite. BSc thesis, University of Bath.

Anderson, K. E., Linksvayer, T., and Smith, C. R. 2008. The causes and consequences of genetic caste determination in ants (Hymenoptera: Formicidae). *Myrmecological News* 11: 119–132.

Anjos, D. V., Luna, P., Borges, C. C., Dattilo, W., and Del-Claro, K. 2019. Structural changes over time in individual-based networks involving harvester ants, seeds, and invertebrates. *Ecological Entomology* 44: 753–761.

Anon. 1950. A secret remedy for cancer. *British Medical Journal* 1950(2): 663–664.

Attewell, P. J., and Wagner, H. C. 2019. *Tetramorium impurum* Foerster (Hymenoptera: Formicidae), first record for Guernsey and the Channel Islands. *British Journal of Entomology and Natural History* 32: 287–295.

Attewell, P. J., Collingwood, C. A., and Godfrey, A. 2010. *Ponera testacea* (Emery, 1895) (Hym.: Formicidae) new to Britain from Dungeness, East Kent. *Entomologist's Record and Journal of Variation* 122: 113–119.

Banks, W. A., Adams, C. T., and Lofgren, C. S. 1991. Damage to young citrus trees by red imported fire ant (Hymenoptera: Formicidae). *Journal of Economic Entomology* 84: 241–246.

Barden, P. 2017. Fossil ants (Hymenoptera: Formicidae): ancient diversity and the rise of modern lineages. *Myrmecological News* 24: 1–30.

Barnard, P. C. 2011. *The Royal Entomological Society Book of British Insects*. Oxford: Wiley-Blackwell.

Barrett, K. E. J. 1979. *Provisional Atlas of the Insects of the British Isles. Part 5, Hymenoptera: Formicidae, Ants*. Huntingdon: Biological Records Centre.

Bates, H. W. 1863. *The Naturalist on the River Amazons*. London: John Murray.

Baxter, F. P., and Hole, F. D. 1966. Ant (*Formica cinerea*) pedoturbation in a prairie soil. *Soil Science Society of America Proceedings* 31: 425–428.

Beavis, I. C. 1988. *Insects and Other Invertebrates in Classical Antiquity*. Exeter: Exeter University Press.

Bernard, F. 1968. *Les fourmis (Hymenoptera Formicidae) d'Europe occidentale et septentrionale. Faune de l'Europe et du Bassin Méditerranéen 3*. Paris: Masson.

Bernhard, T., and Loe, T. 2015. *Collecting the New Naturalists*. London: Collins.

Bertelsmeier, C., Blight, O., and Courchamp, F. 2016. Invasions of ants (Hymenoptera: Formicidae) in light of global climate change. *Myrmecological News* 22: 25–42.

Bhatkar, A., and Whitcomb, W. H. 1970. Artificial diet for rearing various species of ants. *The Florida Entomologist* 53: 229–232.

Bingley, W. 1813. *Animal Biography, or popular zoology comprising authentic anecdotes of the economy, habits of life, instincts, and sagacity, of the animal creation. Arranged according to the system of Linnaeus.* 3 vols. 4th edition. London: Rivington.

Blacker, N. C., and Collingwood, C. A. 2002. Some significant new records of ants (Hymenoptera: Formicidae) from the Salisbury area, South Wiltshire, England, with a key to the British species of *Lasius. British Journal of Entomology and Natural History* 15: 25–46.

Blaikie, W. G. 1880. *The Personal Life of David Livingstone, chiefly from his unpublished journals and correspondence in the provision of his family.* London: John Murray.

Blatrix, R., Galkowski, C., Lebas, C., and Wegnez, P. 2013. *Fourmis de France, de Belgique et du Luxembourg.* Lonay: Delachaux & Niestlé.

Bliss, P., Katzerke, A., and Neumann, P. 2006. The role of molehills and grasses for filial nest founding in the wood ant *Formica exsecta* (Hymenoptera: Formicidae). *Sociobiology* 47: 903–913.

Bodenheimer, F. S. 1947. The manna of Sinai. *The Biblical Archaeologist* 10: 1–6.

Bold, T. J. 1866. *Myrmica lobicornis* in Durham and Northumberland. *Entomologist's Monthly Magazine* 2: 234.

Bolton, B. 2005. Key to British *Myrmica* species (workers). *BWARS Newsletter* Spring 2005: 13–14.

Bolton, B., and Collingwood, C. A. 1975. *Hymenoptera: Formicidae. Handbooks for the Identification of British Insects 6:3(c).* London: Royal Entomological Society.

Bonnet, C. 1779. Observation LXIII. Sur un procédé des fourmis. In *Oevres d'Histoire Naturelle et de Philosophie.* vol. I, pp. 535–536. Neuchatel: Samuel Fauche.

Boomsma, J. J., and Leusink, A. 1981. Weather conditions during nuptial flights of four European ant species. *Oecologia* 50: 236–241.

Brady, S. G. 2003. Evolution of the army ant syndrome: the origin and long-term evolutionary stasis of a complex of behavioral and reproductive adaptations. *Proceedings of the National Academy of Sciences of the United States of America* 100: 6575–6579.

Brian, M. V. 1977. *Ants.* New Naturalist 59. London: Collins.

Brinker, P., Weig, A., Rambold, G., Feldhaar, H., and Tragust, S. 2018. Microbial community composition of nest-carton and adjoining soil of the ant *Lasius fuliginosus* and the role of host secretions in structuring microbial communities. *Fungal Ecology* 38: 1–10.

Brown, W. L., and Nutting, W. L. 1949. Wing venation and the phylogeny of the Formicidae (Hymenoptera). *Transactions of the American Entomological Society* 75: 113–132.

Brunner, J. 1968. *Stand on Zanzibar.* New York: Doubleday.

Buckley, A. 1880. *Life and Her Children: Glimpses of Animal Life from the Amoeba to the Insects.* London: Stanford.

Budgen, L. M. 1850. *Episodes of Insect Life.* London: Reeve & Benham.

Buschinger, A. 2009. Social parasitism among ants: a review (Hymenoptera: Formicidae). *Myrmecological News* 12: 219–235.

Cardoso, D. C., Santos, H. G., and Cristiano, M. P. 2018. The ant chromosome database – ACdb: an online resource for ant (Hymenoptera: Formicidae) chromosome researchers. *Myrmecological News* 27: 87–91.

Cassill, D. L., and Tschinkel, W. R. 1999. Regulation of diet in the fire ant, *Solenopsis invicta. Journal of Insect Behaviour* 12: 307–328.

Cassill, D. L., Butler, J., Vinson, S. B., and Wheeler, D. E. 2005. Cooperation during prey digestion between workers and larvae in the ant, *Pheidole spadonia. Insectes Sociaux* 52: 339–343.

Champion, G. C. 1870. Captures of Coleoptera during the last season. *Entomologist's Monthly Magazine* 6: 231–232.

Chapman, J. W., Reynolds, D. R., Smith, A. D., Smith, E. T., and Woiwod, I. P. 2004. An aerial netting study of insects migrating at high altitude over England. *Bulletin of Entomological Research* 94: 123–136.

Chapman, T. A. 1916. What the larva of *Lycaena arion* does during its last instar. *Transactions of the Royal Entomological Society of London* 1915: 291–297.

Cherrett, J. M. 1986. History of the leaf-cutting ant problem. In: Lofgren, C. S., and Van der Meer, R. K. eds, *Fire Ants and Leaf-Cutting Ants: Biology and Management*. Boulder: Westview Press.

Chick, L. D., Perez, A., and Diamond, S. E. 2017. Social dimensions of physiological responses to global climate change: what we can learn from ants (Hymenoptera: Formicidae). *Myrmecological News* 25: 29–40.

Collingwood, C. A. 1962. *Myrmica puerilis* Staercke, 1942, an ant new to Britain. *Entomologist's Monthly Magazine* 98: 18–20.

Collingwood, C. A. 1979. *The Formicidae (Hymenoptera) of Fennoscandia and Denmark*. Fauna Entomologica Scandinavica 8. Klampenborg: Scandinavian Science Press.

Collins, C. M., and Leather, S. 2002. Ant-mediated dispersal of the black willow aphid *Pterocomma salicis* L.; does the ant *Lasius niger* judge aphid-host quality? *Ecological Entomology* 27: 238–241.

Comstock, J. H., and Needham, J. G. 1898–1899. The wings of insects. *American Naturalist* 32: 43–48, 81–89, 231–257, 335–340; 33: 117–126, 573–582, 845–860.

Cooke, D., Norriss, T., Fomison, L., and Plant, C. W. 2013. *Myrmeleon formicarius* (L., 1767) (Neur.: Myrmeleontidae): an ant-lion new to Britain. *Entomologist's Record and Journal of Variation* 125: 174–178.

Crawley, W. C. 1920. A new species of ant imported into England. *Entomologist's Record and Journal of Variation* 32: 180–181.

Crozier, R. H., Jermin, L. S., and Chiotis, M. 1997. Molecular evidence for the Jurassic origin of ants. *Naturwissenschaften* 84: 22–23.

Csata, E., and Dussutour, A. 2019. Nutrient regulation in ants (Hymenoptera: Formicidae): a review. *Myrmecological News* 29: 111–124.

Curtis, J. 1823–1840. *British Entomology: being illustrations and descriptions of the genera of insects found in Great Britain and Ireland: containing coloured figures from nature of the most rare and beautiful species, and in many instances of the plants upon which they are found*. 8 vols. London: Printed for the author.

Curtis, J. 1854. On the genus *Myrmica*, and other indigenous ants. *Transactions of the Linnean Society of London* 21: 211–220.

Czaczkes, T. J., Grüter, C., and Ratnieks, L. W. 2015. Trail pheromones: an integrative view of their role in insect colony organisation. *Annual Review of Entomology* 60: 581–599.

Czechowski, W., Radchenko, A., and Czechowska, W. 2002. *The Ants (Hymenoptera, Formicidae) of Poland*. Warsaw: Museum and Institute of Zoology.

Czechowski, W., Radchenko, A., Czechowska, W., and Vespalainen, K. 2012. *The Ants of Poland with Reference to the Myrmecofauna of Europe*. Fauna Poloniae (New Series) 4. Warsaw: Natura optima dux Foundation.

Dandy, J. E. 1969. *Watsonian Vice-Counties of Great Britain*. London: Ray Society.

Darwin, C. 1871. *The Descent of Man and Selection Relative to Sex*. London: John Murray.

Dawkins, R. 1976. *The Selfish Gene*. Oxford: Oxford University Press.

DeFoliart, G. R. 1999. Insects as food: why the western attitude is important. *Annual Review of Entomology* 44: 21–50.

DeLucia, E. H., Nabity, P. D., Zavala, J. A., and Berenbaum, M. R. 2012. Climate change: resetting plant–insect interactions. *Plant Physiology* 160: 1677–1685.

Denton, J. 2018. *Lasius brunneus* (Letreille) (Hymenoptera: Formicidae) in East Kent and South Hampshire. *British Journal of Entomology and Natural History* 31: 42.

Denton, J., and Collins, G. A. 2004. The black bog ant *Formica candida* (Smith) (Hymenoptera: Formicidae) in Surrey. *British Journal of Entomology and Natural History* 17: 110.

de Oliveira, J. F., Wajnberg, E., de Souza Esquivel, D. M., Weinkauf, S., and Winklhofer, M. 2009. Ant antennae: are they sites for magnetoreception? *Journal of the Royal Society Interface* 7: 143–152.

de Vega, C., Arista, M., Ortiz, P. L., Herrera, C. M., and Talavera, S. 2009. The ant-pollination system of *Cytinus hypocistis* (Cytinaceae), a Mediterranean root holoparasite. *Annals of Botany* 103: 1065–1075.

De Vita, J. 1979. Mechanisms of interference and foraging among colonies of the harvester ant *Pogonomyrmex californicus* in the Mojave Desert. *Ecology* 60: 729–737.

Di Giulio, A., Maurizi, E., Barbero, F., Sala, M., Fattorini, S., Balletto, E., and Bonelli, S. 2015. The Pied Piper: a parasitic beetle's melodies modulate ant behaviours. *PLoS ONE* 10(7): e0130541. doi:

10.1371/journal.pone.0130541.

Domisch, T., Finér, L., Neuvonen, S., Niemelä, P., Risch, A. C., Kilpeläinen, J., Ohashi, M., and Jurgensen, M. F. 2009. Foraging activity and dietary spectrum of wood ants (*Formica rufa* group) and their role in nutrient fluxes in boreal forests. *Ecological Entomology* 34: 369–377.

Donisthorpe, H. St J. K. 1915. *British Ants: Their Life History and Classification*. Plymouth: W. Brendon & Son.

Donisthorpe, H. 1923. *Acanthomyops (Donisthorpea) brunneus* Latr., a species of Formicidae new to Britain. *Entomologist's Record and Journal of Variation* 35: 21–23.

Donisthorpe, H. St J. K. 1927a. *British Ants: Their Life History and Classification*, 2nd edition. London: Routledge.

Donisthorpe, H. St J. K. 1927b. *The Guests of British Ants: Their Habits and Life-Histories*. London: Routledge.

Donisthorpe, H. St J. K. 1939. *A Preliminary List of the Coleoptera of Windsor Forest*. London: Lloyd.

Dubois, M. B. 1993. What's in a name? A clarification of *Stenamma westwoodii, S. debile*, and *S. lippulum* (Hymenoptera: Formicidae: Myrmicinae). *Sociobiology* 21: 299–334.

Dubois, M. B. 1998. A revision of the ant genus *Stenamma* in the Palaearctic and Oriental regions. *Sociobiology* 32: 193–403.

Duncan, P. M. 1883. *Cassell's Natural History*. London: Cassell.

Eaton, A. E. 1869. An early swarm of *Formica nigra*. *Entomologist's Monthly Magazine* 5: 298.

Ebsen, J. R., Boomsma, J. J., and Nash, D. R. 2019. Phylogeography and cryptic speciation in the *Myrmica scabrinodis* Nylander, 1846 species complex (Hymenoptera: Formicidae), and their conservation implications. *Insect Conservation and Diversity* 12: 467–480.

Edwards, J. P., and Baker, L. F. 1981. Distribution and importance of the pharaoh's ant *Monomorium pharaonis* (L.) in National Health Service hospitals in England. *Journal of Hospital Infection* 2: 249–254.

Elmes, G. W. 1978. A morphometric comparison of three closely related species of *Myrmica* (Formicidae), including a new species from England. *Systematic Entomology* 3: 131–145.

Elmes, G. W., Radchenko, A. G., and Thomas, J. A. 2003. First records of *Myrmica vandeli* Bondroit

(Hymenoptera, Formicidae) for Britain. *British Journal of Entomology and Natural History* 16: 145–152.

El-Ziady, S., and Kennedy, J. S. 1956. Beneficial effects of the common garden ant, *Lasius niger* L., on the black bean aphid, *Aphis fabae* Scopoli. *Proceedings of the Royal Entomological Society London (A)* 31: 61–65.

Falk, S. 1991. A review of the scarce and threatened bees, wasps and ants of Great Britain. Peterborough: Nature Conservancy Council.

Ferenczy, A. 1924. *The Ants of Timothy Thümmel*. London: Jonathan Cape.

Finnegan, R. J. 1975. Introduction of the predaceous red wood ant, *Formica lugubris* (Hymenoptera: Formicidae), from Italy into Eastern Canada. *Canadian Entomologist* 107: 1721–1274.

Fitch, A. 1856. First and second report on the noxious, beneficial and other insects of the state of New York, made to the State Agricultural Society, pursuant to an appropriation for the purpose from the legislature of the state. Albany: Van Benthuysen.

Fittkau, E. J., and Klinge, H. 1973. On biomass and trophic structure of the central Amazonian rain forest ecosystem. *Biotropica* 5: 2–14.

Fitton, M. G., Graham, M. W. R. de V., Boucek, Z. R. J., Fergusson, N. D. M., Huddleston, T., Quinlan, J., and Richards, O. W. 1978. *A Check List of British insects. Part 4. Hymenoptera*. Handbooks for the Identification of British Insects (11:4). London: Royal Entomological Society.

Forbes, S. A. 1906. The corn root-aphis and its attendant ant (*Aphis maidiradicis* Forbes and *Lasius niger* L. var *americanus* Emery). *Bulletin of the United States Department of Agriculture, Division of Entomology* 60: 29–39.

Forel, A. 1874. *Les fourmis de la Suisse*. Zurich: Société Helvétique des Sciences Naturelles.

Forel, A. 1886. Etudes myrmécologiques en 1886. *Annales de la Société Entomologique de Belgique* 30: 131–215.

Forel, A. 1921–1923. *Le monde social des fourmis du globe comparé à celui de l'homme*. 5 volumes. Geneva: Kundig.

Fox, M. G. 2010. First incursion of *Lasius neglectus* (Hymenoptera: Formicidae), an invasive polygynous ant in Britain. *British Journal of Entomology and Natural History* 23: 259–261.

Fox, M. 2011. *Temnothorax unifasciatus* and *T. tuberum* in Britain. *BWARS Newsletter* Autumn 2011: 24.

Fox, M., and Wang, C. 2016. Colony of the Argentine ant, *Linepithema humile* (Hymenoptera: Formicidae), in Fulham, West London. *British Journal of Entomology and Natural History* 29: 193–195.

Frank, E. T., Wehrhahn, M., and Linssenmair, K. E. 2018. Wound treatment and selective help in a termite-hunting ant. *Proceedings of the Royal Society B* 285: 20172457. doi: 10.1098/rspb.2017.2457.

Frohawk, F. W. 1916. Further observations on the last stage of the larva of *Maculinea arion*. *Transactions of the Royal Entomological Society of London* 1915: 313–316.

Garcia, L. C., and Styrsky, J. D. 2013. An orb-weaver spider eludes plant-defending acacia ants by hiding in plain sight. *Ecological Entomology* 38: 230–237.

Garrido, J. L., Rey, P. J., Cerda, X., and Herrera, C. M. 2002. Geographical variation in diaspore traits of an ant-dispersed plant (*Helleborus foetidus*): are ant community composition and diaspore traits correlated? *Journal of Ecology* 90: 446–455.

Gauld, D., and Bolton, B. 1988. *The Hymenoptera*. London: British Museum (Natural History).

Giraud, T., Pederson, J. S. S., and Keller, L. 2002. Evolution of supercolonies: the Argentine ants of southern Europe. *Proceedings of the National Academy of Sciences of the United States of America* 99: 6075–6079.

Goedart, J., and Lister, M. 1685. *De insectis, in methodum redactus; cum notularum additione. Metamorphosis Naturalis*. London: Smith.

Goidanich, A. 1959. Le migrazioni coatte mirmecogene dello *Stomaphis quercus* Linnaeus, afido olociciclio monoico omotopo. *Bollettino dell'Istituto di Entomologia della Università degli Studi di Bologna* 23: 93–131.

Goldenberg, I. S. 1959. Catgut, silk and silver – the story of surgical sutures. *Surgery* 5: 905–912.

Gomes, I. J. M. T., Santiago, D. F., Campos, R. I., and Vasconcelos, H. L. 2019. Why do *Pheidole oxyops* (Forel, 1908) ants place feathers around their nests? *Ecological Entomology* 44: 451–456.

Gould, W. 1747. *An Account of English Ants; which contains I. Their different species and mechanism. II. Their manner of government, and a description of their several queens. III. The production of their eggs and process of the young. IV. The incessant labours of the workers or common ants. With many other curiosities observable in these surprising insects.* London: Millar.

Graedel, T. E., and Eisner, T. 1988. Atmospheric formic acid from formicine ants: a preliminary assessment. *Tellus*, series B 40: 335–339.

Grasso, D. A., Giannetti, D., Castracani, C., Spotti, F. A., and Mori, A. 2020. Rolling away: a novel context-dependent behaviour discovered in ants. *Scientific Reports* 10: 3784. doi: 10.1038/s41598-020-59954-9.

Grimaldi, D., and Engel, M. S. 2005. *Evolution of the Insects*. Cambridge: Cambridge University Press.

Guénard, B., Weiser, M. D., Gómez, K., Narula, N., and Economo, E. P. 2017. The Global Ant Biodiversity Informatics (GABI) database: synthesizing data on the geographic distribution of ant species (Hymenoptera: Formicidae). *Myrmecological News* 24: 83–89.

Haines, B. L. 1978. Element and energy flows through colonies of the leafcutter ant, *Atta colombica*, in Panama. *Biotropica* 10: 270–277.

Halstead, A. J. 2001. [Exhibit to BENHS meeting, 12 June 2001, of unnamed 'exotic black ant' found on date palms, *Phoenix dactylifera*, Chelsea Flower Show, May 2001.] *British Journal of Entomology and Natural History* 14: 236.

Hamer, M. T., and Cocks, L. R. 2020. *Linepithema iniquum* (Mayr) (Hymenoptera: Formicidae) found at the National Botanic Garden of Wales. *British Journal of Entomology and Natural History* 33: 71–75.

Hamer, M. T., Marquis, A. D., and Guénard, B. 2021. *Strumigenys perplexa* (Smith, 1876) (Formicidae, Myrmicinae) a new exotic ant to Europe with establishment in Guernsey, Channel Islands. *Journal of Hymenoptera Research* 83: 101–124.

Hamilton, W. D. 1964. The genetical evolution of social behaviour. *Journal of Theoretical Biology* 7: 1–16, 17–52.

Hancock, E. G. and Robinson, J. 2021. Alluaud's little yellow ant, *Plagiolepis alluaudi* (Hymenoptera: Formicidae) in Scotland. *British Journal of Entomology and Natural History* 34: 69-75.

Hangartner, W. 1967. Spezifität und Inaktivierung des Spurpheromons von *Lasius fuliginosus* Latr. und Orientierung der Arbeiterinnen in Duftfeld. *Zeitschrift für Vergleichende Physiologie* 57: 103–136.

Hart, A. G., Hesselberg, T., Nesbit, R., and

Goodenough, A. E. 2018. The spatial distribution and environmental triggers of ant mating flights: using citizen-science data to reveal national patterns. *Ecography* 41: 877–888.

Hawkins, T. H. 1953. *The Ant World* [book review]. *Nature* 172: 473.

Heads, P. A., and Lawton, J. H. 1984. Bracken, ants and extrafloral nectaries. II. The effect of ants on the insect herbivores of bracken. *Journal of Animal Ecology* 53: 1015–1031.

Heinze, J. 2008. The demise of the standard ant (Hymenoptera: Formicidae). *Myrmecological News* 11: 9–20.

Heinze, J., and Rueppell, O. 2014. The frequency of multi-queen colonies increases with altitude in a Nearctic ant. *Ecological Entomology* 39: 527–529.

Helms, J. A. 2018. The flight ecology of ants (Hymenoptera: Formicidae). *Myrmecological News* 26: 19–30.

Helms, J. A., Godfrey, A. P., Ames, T., and Bridge, E. S. 2016. Are invasive fire ants kept in check by native aerial insectivores? *Biology Letters* 12: 20160059. doi: 10.1098/rsbl.2016.0059.

Heyes, B. 1993. Little hills of cushioned thyme. *The John Clare Society Journal* 12: 32–36.

Hickman, J. C. 1974. Pollination by ants – a low-energy system. *Science* 184: 1290–1292.

Higashi, S., and Yamauchi, K. 1979. Influence of a supercolonial ant *Formica* (*Formica*) *yessensis* Forel on the distribution of other ants on Ishikari Coast. *Japanese Journal of Ecology* 29: 257–264.

Hill, J. K., Rosengaus, R. B., Gilbert, F. S., and Hart, A. G. 2013. Invasive ants – are fire ants drivers of biodiversity loss? *Ecological Entomology* 38: 539.

Hölldobler, B., and Kwapich, C. L. 2017. *Amphotis marginata* (Coleoptera: Nitidulidae) a highwayman of the ant *Lasius fuliginosus*. *PLoS ONE* 12(8): e0180847. doi: 10.1371/journal.pone.0180847.

Hölldobler, B., and Wilson, E. O. 1990. *The Ants*. Cambridge, MA: Belknap Press.

Hölldobler, B., and Wilson, E. O. 1994. *Journey to the Ants*. Cambridge, MA: Harvard University Press.

Hölldobler, B., and Wilson, E. O. 2009. *The Superorganism: the Beauty, Elegance and Strangeness of Insect Societies*. New York: Norton.

Howard, L. O. 1904. *The Insect Book: a popular account of the bees, wasps, ants, grasshoppers, flies, and other North American insects exclusive of butterflies, moths and beetles, with full life histories, tables and bibliographies.* Garden City: Doubleday.

Hurlbert, A. H., Ballantyne, F. IV, and Powell, S. 2008. Shaking a leg and hot to trot: the effects of body size and temperature on running speed in ants. *Ecological Entomology* 33: 144–154.

Ivens, A. B. F., Kronauer, D. J. C., Weissing, F. J., and Boomsma, J. J. 2012. Ants farm subterranean aphids mostly in single clone groups – an example of prudent husbandry for carbohydrates and proteins? *BioMed Central Evolutionary Biology* 12: 106.

Jolivet, P. 1998. *Interrelationship Between Insects and Plants*. Boca Raton, FL: CRC Press.

Jones, C. R. 1927. Ants and their relation to aphids. PhD thesis, Iowa State College, Ames.

Jones, R. A. 1984. [Ants found in a box of ants' eggs, exhibited at indoor meeting of 22 March 1984.] *Proceedings and Transactions of the British Entomological and Natural History Society* 17: 99.

Jones, R. A. 1994. Ants feeding directly on plant sap. *British Journal of Entomology and Natural History* 7: 139–140.

Jones, R. A. 1996. *Lasius brunneus* (Latreille) (Hymenoptera: Formicidae) new to Kent? *British Journal of Entomology and Natural History* 9: 135–136.

Jones, R. A. 1997. Life on the edge – a caution on the precise demarcation of Watsonian vice-county boundaries in the London area. *London Naturalist* 76: 79–81.

Jones, R. A. 1998. *Hypoponera punctatissima* (Roger) (Hymenoptera: Formicidae) in south-east London. *British Journal of Entomology and Natural History* 10 (1997): 256.

Jones, R. A. 2003. *Lasius brunneus* (Latr.) (Hymenoptera: Formicidae) found indoors. *British Journal of Entomology and Natural History* 16: 219.

Jones, R. A. 2004. Green deserts? The invertebrate fauna of mown grass playing fields. *British Journal of Entomology and Natural History* 17: 39–44.

Jones, R. A. 2010. *Extreme Insects*. London: HarperCollins.

Jones, R. A. 2013. Aerial plankton and a mass drowning of insects in northern France. *British Journal of Entomology and Natural History* 26: 9–15.

Jones, R. A. 2015. *House Guests – House Pests: a Natural History of the Animals in Our Homes*. London: Bloomsbury.

Jones, R. A. 2016. *Temnothorax nylanderi* (Foerster)

(Hymenoptera: Formicidae) nesting in cramp balls. *British Journal of Entomology and Natural History* 29: 96.

Jones, R. A., and Ure-Jones, C. 2021. *A Natural History of Insects in 100 Limericks.* Exeter: Pelagic Publishing.

Jurgensen, M. F., Storer, A. J., and Risch, A. C. 2005. Red wood ants in North America. *Annales Zoologici Fennici* 42: 235–242.

Keller, L., and Ross, K. 1998. Selfish genes: a green beard in the red fire ant. *Nature* 394: 573–575.

Kennedy, P., Higginson, A. D., Radford, A. N., and Sumner, S. 2018. Altruism in a volatile world. *Nature* 555: 359–362.

King, J. R., and Tschinkel, W. R. 2013a. Experimental evidence for weak effects of fire ants in a naturally invaded pine-savanna ecosystem in north Florida. *Ecological Entomology* 38: 68–75.

King, J. R., and Tschinkel, W. R. 2013b. Fire ants are not drivers of biodiversity change: a response to Stuble *et al.* (2013). *Ecological Entomology* 38: 543–545.

King, T. J. 1977. The plant ecology of ant-hills in calcareous grasslands. *Journal of Ecology* 65: 235–256, 257–278, 279–315.

Kirby, W., and Spence, W. 1818. *An Introduction to Entomology: or elements of the natural history of insects.* Volume 2. London: Longman, Hurst, Rees, Orme & Brown.

Kloet, G. S., and Hincks, W. D. 1945. *A Check List of British Insects.* Stockport: privately printed.

Kloft, W. J. 1959. Versuch einer Analyse der trophobiotische Beziehungen von Ameisen und Aphiden. *Biologisches Zentralblatt* 78: 863–870.

Kovac, H., Stabentheiner, A., Hetz, S. K., Petz, M., and Crailsham, K. 2007. Respiration of resting honeybees. *Journal of Insect Physiology* 53: 1250–1261.

Kramer, B. H., Schaible, R., and Scheuerlein, A. 2016. Worker lifespan is an adaptive trait during colony establishment in the long-lived ant *Lasius niger*. *Experimental Gerontology* 85: 18–23.

Kukalová-Peck, J. 1978. Origin and evolution of insect wings and their relation to metamorphosis, as documented by the fossil record. *Journal of Morphology* 156: 53–126.

Kukalová-Peck, J. 1983. Origin of the insect wing and articulation from the arthropodan leg. *Canadian Journal of Zoology* 71: 1618–1669.

Kutter, H. 1969. Die sozialparasitischen Ameisen der Schweiz. *Neujahrsblatt der Naturforsch Gesellschaft Zurich* 113: 1–62.

Kutter, H., and Stumper, R. 1969. Hermann Appel, ein leidgeadelter Entomologie (1892–1966). In *Proceedings of the Sixth International Congress of the International Union for the Study of Social Insects*, pp. 275–279. Bern: University of Bern Zoological Institute.

Landsell, H. 1908. *The Tithe in Scripture, being chapters from 'The sacred tenth' with a revised bibliography on tithe-paying and systematic and proportionate giving.* London: Society for Promoting Christian Knowledge.

LaPolla, J. S., Dlussky, G. M., and Perrichot, V. 2013. Ants in the fossil record. *Annual Review of Entomology* 58: 609–630.

Latreille, P. A. 1802. *Histoire naturelle des fourmis: et recueil de mémoires et d'observations sur les abeilles, les araignées, les faucheurs, et autres insects.* Paris: Barrois.

Leather, S. R. 2015. An entomological classic – the pooter or insect aspirator. *British Journal of Entomology and Natural History* 28: 52–54.

Leather, S. 2017. Not all aphids are farmed by ants. https://simonleather.wordpress.com/2017/01/30/not-all-aphids-are-farmed-by-ants (accessed February 2021).

Lebas, C., Galkowski, C., Blatrix, R., and Wegnez, P. 2019. *Ants of Britain and Europe: a Photographic Guide.* London: Bloomsbury Wildlife.

Lescano, M. N., Farji-Brener, A. G., and Gianoli, E. 2014. Nocturnal resource defence in aphid-tending ants of northern Patagonia. *Ecological Entomology* 39: 203–209.

Lush, M. J. 2007. *Lasius brunneus* (Latreille) (Hymenoptera: Formicidae) in Somerset. *British Journal of Entomology and Natural History* 20: 74.

Macdonald, M. 2013. *Highland Ants: Distribution, Ecology and Conservation.* Inverness: Highland Biological Recording Group.

MacKay, W., and MacKay, E. 1985. Temperature modifications of the nest of *Pogonomyrmex montanus* (Hymenoptera: Formicidae). *The Southwestern Naturalist* 30: 307–309.

Mallumphy, C. 2016. The status of Alluaud's little yellow ant *Plagiolepis alluaudi* Emery (Hymenoptera: Formicidae) in Britain. *British Journal of Entomology and Natural History* 29: 65–68.

Mansor, M. S., Abdullah, N. A., Halim, M. R. A.,

Nor, S. M., and Ramli, R. 2018. Diet of tropical insectivorous birds in lowland Malaysian rainforest. *Journal of Natural History* 52: 2301–2316.

Markin, G. P., Dillier, J. H., Hill, S. O., Blum, M. S., and Hermann, H. R. 1971. Nuptial flight and flight ranges of the imported fire ant *Solenopsis saevissima richteri* (Hymenoptera: Formicidae). *Journal of the Georgia Entomological Society* 6: 145–159.

Maruyama, M., and Parker, J. 2017. Deep-time convergence in rove beetle symbionts of army ants. *Current Biology* 27: 920–926.

Matthews, J., and Matthews, C. 2005. *The Element Encyclopedia of Magical Creatures. The Ultimate A–Z of Fantastic Beings in Myth and Legend.* London: Harper Element.

Mawdsley, J. R. 1994. Mimicry and Cleridae (Coleoptera). *Coleopterists Bulletin* 48: 115–125.

Mayer, V. E., and Vogelmayr, H. 2009. Mycelial carton galleries of *Azteca brevis* (Formicidae) as a multi-species network. *Proceedings of the Royal Society B* 276: 3265–3273.

McClenaghan, I. 2006. *Myrmecina graminicola* on apples on the ground. *BWARS Newsletter* Autumn 2006: 22.

Merian, M. S. 1705. *Metamorphosis insectorum surinamensium, ofte verandering der surinaamsche insecten.* Amsterdam.

Michelet, J. 1875. *The Insect.* London: Nelson.

Mittler, T. E. 1958. Studies on the feeding and nutrition of *Tuberolachnus salignus* (Gmelin) (Homoptera, Aphididae). II. The nitrogen and sugar composition of ingested phloem sap and excreted honeydew. *Journal of Experimental Biology* 35: 74–84.

Moggridge, J. T. 1873. *Harvesting Ants and Trap-Door Spiders. Notes and observations on their habits and dwellings.* London: Reeve.

Möller, A. 1893. Die Pilzgärten einiger südamerikaischer Ameisen. In: Schimper A. F. W. ed., *Botanische Mittheilungen aus den Tropen*, vol. 6. Jena: Fischer.

Moreau, C. S. 2009. Inferring ant evolution in the age of molecular data (Hymenoptera: Formicidae). *Myrmecological News* 12: 201–210.

Morgan, E. D. 2008. Chemical sorcery for sociality: exocrine secretion of ants (Hymenoptera: Formicidae). *Myrmecological News* 11: 79–90.

Morgan, E. D. 2009. Trail pheromones of ants. *Physiological Entomology* 34: 1–17.

Morice, F. D., and Durrant, J. H. 1915. The authorship and first publication of the 'Jurinean' genera of Hymenoptera: being a reprint of a long-lost work by Panzer, with a translation into English, an introduction, and bibliographical and critical notes. *Transactions of the Royal Entomological Society of London* 1914: 421–423.

Morley, B. D. W. 1939. Ants' ninth sense – one of the mysteries of ant life. *Journal of the Transactions of the Victoria Institute or Philosophical Society of Great Britain* 61: 80–98.

Morley, C. 1909. *A Guide to the Natural History of the Isle of Wight.* Newport: County Press.

Morley, D. W. 1953a. *The Ant World: Their Evolutionary History, Their Behaviour, Their Many Varieties, and the Organisation of Their Society.* London: Penguin Books.

Morley, D. W. 1953b. *Ants.* New Naturalist Monograph 8. London: Collins.

Morozov, N. S. 2014. Why do birds practice anting? *Biology Bulletin Reviews* 5: 353–365.

Mothapo, N. P., and Wossler, T. C. 2014. Resource competition assays between the African big-headed ant, *Pheidole megacephala* (Fabricius) and the invasive Argentine ant, *Linepithema humile* (Mayr): mechanisms of inter-specific displacement. *Ecological Entomology* 39: 501–510.

Mouffet, T. 1658. *The Theater of Insects: or lesser living creatures as bees, flies, caterpillars, spidrs* [sic], *worms, & co, a most elaborate work.* London: Cotes.

Nalepa, C. A. 2015. Origin of termite eusociality: trophallaxis integrates the social, nutritional, and microbial environments. *Ecological Entomology* 40: 323–335.

Newman, E. 1841. *A Familiar Introduction to the History of Insects; being a new and greatly improved edition of The grammar of entomology.* London: Van Voorst.

Orledge, G. M. 2005. The BWARS *Lasius* study: differences between the workers of *L. niger* and *L. platythorax* illustrated with scanning electron micrographs. *BWARS Newsletter* Autumn 2005: 22–24.

Orlowski, G., and Karg, J. 2013. Diet breadth and overlap in three sympatric aerial insectivorous birds at the same location. *Bird Study* 60: 475–483.

Orwell, G. 1944. Benefit of clergy: some notes on Salvador Dali. In: *The Saturday Book for 1944.* London: Hutchinson.

Parker, J. 2016. Myrmecophily in beetles (Coleoptera): evolutionary patterns and biological mechanisms. *Myrmecological News* 22: 65–108.

Parker, J., and Grimaldi, D. A. 2014. Specialized myrmecophily at the ecological dawn of modern ants. *Current Biology* 24: 2428–2434.

Parmentier, T., Dekoninck, W., and Wenseleers, T. 2014. A highly diverse microcosm in a hostile world: a review of the associates of red wood ants (Formica rufa group). *Insectes Sociaux* 61: 229–237. doi: 10.1007/s00040-014-0357-3.

Patek, S. N., Baio, J. E., Fisher, B. L., and Suarez, A. V. 2006. Multifunctionality and mechanical origins: ballistic jaw propulsion in trap-jaw ants. *Proceedings of the National Academy of Sciences of the United States of America* 103: 12787–12792.

Paul, J. 2011. *Ponera testacea* Emery, 1895 (Hymenoptera: Formicidae) present at multiple sites along the south coast of England. *BWARS Newsletter* Autumn 2011: 24–25.

Peakall, R., Beattie, A. J., and James, S. H. 1987. Pseudocopulation of an orchid by male ants: a test of two hypotheses accounting for the rarity of ant pollination. *Oecologia* 73: 522–524.

Perfilieva, K. S. 2000. Wing venation anomalies in sexual individuals of ants (Hymenoptera, Formicidae) with different strategies of mating behaviour. *Entomological Review* 79: 100–200.

Perfilieva, K. S. 2010. Trends in evolution of ant wing venation (Hymenoptera, Formicidae). *Entomological Review* 90: 857–870.

Perfilieva, K. S. 2015. The evolution of diagnostic characters of wing venation in representatives of the subfamily Myrmeciinae (Hymenoptera, Formicidae). *Entomological Review* 95: 1000–1008.

Perkovsky, E. E. 2011. Synclusions of the Eocene winter ant *Prenolepis henschei* (Hymenoptera: Formicidae) and *Germaraphis aphids* (Hemiptera: Eriosomatidae) in the late Eocene Baltic and Ronvo ambers: some implications. *Russian Entomological Journal* 20: 303–313.

Perrichot, V., Nel, A., Néraudeau, D., Lacau, S., and Guyot, T. 2008. New fossil ants in French Cretaceous amber (Hymenoptera: Formicidae). *Naturwissenschaften* 95: 91–97.

Pfeffer, S. E., Wahl, V. L., Wittlinger, M., and Wolf, H. 2019. High-speed locomotion in the Saharan silver ant, *Cataglyphis bombycina. Journal of Experimental Biology* 222: jeb198705. doi: 10.1242/jeb.198705.

Picker, M. D., Ross-Gillespie, V., Vlieghe, K., and Moll, E. 2012. Ants and the enigmatic Namibian fairy circles – cause and effect? *Ecological Entomology* 37: 33–42.

Pirk, G. I., and Lopez de Casenave, J. 2014. Effect of harvester ants of the genus *Pogonomyrmex* on the soil seed bank around their nests in the central Monte desert, Argentina. *Ecological Entomology* 39: 610–619.

Plant, C. W. 1998. Investigations into the distribution, status and ecology of the ant-lion *Euroleon nostras* (Geoffroy in Fourcroy, 1785) (Neuroptera: Myrmeleontidae) in England during 1997. *Transactions of the Suffolk Naturalists' Society* 34: 69–79.

Poinar, G. O. 1993. Insects in amber. *Annual Review of Entomology* 46: 145–159.

Poinar, G. 2010. Palaeoecological perspectives in Dominican amber. *Annales de la Société Entomologique de France* 46: 23–52.

Pontin, A. J. 1978. The numbers and distribution of subterranean aphids and their exploitation by the ant *Lasius flavus* (Fabr.). *Ecological Entomology* 3: 203–207.

Pontin, J. 2005. *Ants of Surrey.* Pirbright: Surrey Wildlife Trust.

Praeger, R. L. 1901. Irish topographical botany. *Proceedings of the Royal Irish Academy* 7: i–ii, v–clxxxviii, 1–410.

Prior, K. M., Saxena, K., and Frederickson, M. E. 2014. Seed handling behaviours of native and invasive seed-dispersing ants differentially influence seedling emergence in an introduced plant. *Ecological Entomology* 39: 66–74.

Pulliainen, U., Helanterä, H., Sundrstöm, L., and Schultner, E. 2019 The possible role of ant larvae in the defence against social parasites. *Proceedings of the Royal Society B* 286: 20182867. doi: 10.1098/rspb.2018.2867.

Ray, J. 1710. *Historia insectorum.* London: Churchill.

Razeng, E., and Watson, D. M. 2015. Nutritional composition of the preferred prey of insectivorous birds: popularity reflects quality. *Journal of Avian Biology* 46: 89–96.

Reemer, M. 2013. Review and phylogenetic evaluation of associations between Microdontinae (Diptera: Syrphinae) and ants (Hymenoptera: Formicidae). *Psyche* 2013: 538316. doi: 10.1155/2013/538316.

Rennie, J. 1845. *Insect Architecture: to which are*

added, miscellanies on the ravages, the preservation for the purposes of study, and the classification of insects. In two volumes. London: Knight.

Reymond, A., Purcell, J., Cherix, D., Guisan, A., and Pellissier, L. 2013. Functional diversity decreases with temperature in high altitude ant fauna. *Ecological Entomology* 38: 364–373.

Roberts, E. M., and Tapanila, L. 2006. A new social insect nest from the Upper Cretaceous Kaiparowits formation of southern Utah. *Journal of Paleontology* 80: 768–774.

Robinson, N. A. 1999. Observations on the 'guest ant' *Formicoxenus nitidulus* Nylander in nests of the wood ants *Formica rufa* L. and *F. lugubris* Zetterstedt in 1998. *British Journal of Entomology and Natural History* 12: 138–140.

Robinson, N. A. 2001. Changes in the status of the red wood ant *Formica rufa* L. (Hymenoptera: Formicidae) in north west England during the 20th century. *British Journal of Entomology and Natural History* 14: 29–38.

Robinson, N. A., and Woodgate, J. N. 2004. A study of the wood ant *Formica lugubris* Zetterstedt (Hymenoptera: Formicidae) in Ashness Woods, Borrowdale, Cumbria, England in 2001 and 2003. *British Journal of Entomology and Natural History* 17: 203–211.

Rössler, W. 2019. Neuroplasticity in desert ants (Hymenoptera: Formicidae) – importance for the ontogeny of navigation. *Myrmecological News* 29: 1–20.

Rust, J., and Andersen, N. M. 1999. Giant ants from the Paleogene of Denmark with a discussion of the fossil history and early evolution of ants (Hymenoptera: Formicidae). *Zoological Journal of the Linnean Society* 125: 331–348.

Sakata, H. 1994. How an ant decides to prey on or to attend aphids. *Research on Population Ecology* 36: 45–51.

Saunders, E. 1896. *The Hymenoptera Aculeata of the British Islands.* London: Reeve.

Saunders, E. 1902. Hymenoptera aculeata, bees, wasps, ants etc. In *A History of the County of Surrey.* Part 3, Zoology, pp. 84–90. London: Constable.

Schär, S., Talavera, G., Espadaler, X., Rana, J. D., Anderson, A. A., Cover, S. P., and Vila, R. 2018. Do Holarctic ant species exist? Trans-Beringian dispersal and homoplasy in the Formicidae. *Journal of Biogeography* 2018: 1–12.

Schiappa, J., and Van Hee, R. 2012. From ants to staples: history and ideas concerning suturing techniques. *Acta Chirurgica Belgica* 112: 395–402.

Seevers, C. H. 1965. The systematics, evolution and zoogeography of staphylinid beetles associated with army ants (Coleoptera: Staphylinidae). *Fieldiana Zoology* 8: 181–202.

Seifert, B. 2007. *Die Ameisen mittel- und nordeuropas.* Tauer: Lutra Verlags und Vertriebsgesellschaft.

Seifert, B. 2013. *Hypoponera ergatandria* (Forel, 1893) – a cosmopolitan tramp species different from *H. punctatissima* (Roger, 1859) (Hymenoptera: Formicidae). *Soil Organisms* 85: 189–201.

Seifert, B. 2018. *The Ants of Central and North Europe.* Tauer: Lutra Verlags und Vertriebsgesellschaft.

Shaw, G. 1806. *General Zoology or Systematic Natural History. Vol. VI, part II. Insecta.* London: Kearsley.

Simroth, H. 1907. Die Aufklarung der sudafrikanischen Nacktschneckenfauna, auf Grund des von Herrn Dr. L. Schultze mitgebrachten Materials. *Zoologischer Anzeiger* 31: 792–798.

Skinner, G. J., and Allen, G. W. 1996. *Ants.* Naturalists' Handbooks 24. Slough: Richmond Publishing Company.

Sladen, F. W. L. 1908. Hymenoptera Aculeata (ants, wasps and bees). In: *The Victoria history of the county of Kent*, ed. W. Page. Volume 1, pp. 114–121. London: Archibald Constable.

Sleigh, C. 2003. *Ant.* London: Reaktion Books.

Sloggett, J. J., Wood, R. A., and Majerus, M. E. N. 1998. Adaptations of *Coccinella magnifica* Redtenbacher, a myrmecophilous coccinellid, to aggression by wood ants (*Formica rufa* group). I. Adult behavioural adaptation, its ecological context and evolution. *Journal of Insect Behaviour* 11: 889–904.

Smith, F. 1851. *List of the Specimens of British Animals in the Collection of the British Museum. Part VI, Hymenoptera Aculeata.* London: Trustees of the British Museum.

Smith, F. 1865. Notes on British Formicidae. *Entomologist's Monthly Magazine* 1: 28–30.

Smith, M., and Williams, M. A. 2008. *Lasius emarginatus* (Olivier, 1792) (Hymenoptera: Formicidae) confirmed as a British species. *BWARS Newsletter* Autumn 2008: 14–15.

Staab, M., Ohl, M., Zhu, C.-D., and Klein, A.-M. 2014. A unique nest-protection strategy in a new species of spider wasp. *PLoS ONE* 9(7): e101592. doi: 10.1371/journal.pone.0101592.

Step, E. 1924. *The Ant: a Popular Account of the Natural History of Ants of All Countries.* London: Hutchinson.

Stephens, J. F. 1829. *A Systematic Catalogue of British Insects, Being an Attempt to Arrange All the Hitherto Discovered Indigenous Insects in Accordance with Their Natural Affinities.* London: Baldwin & Cradock.

Stockan, J. A., Robinson, E. J. H., and Littlewood, N. A. 2017. Surveys for *Formicoxenus nitidulus* (Nylander) (Hymenoptera: Formicidae) in northern Scotland. *British Journal of Entomology and Natural History* 30: 245–246.

Stubbs, A. E. 1982. Conservation and the future for the field entomologist. *Proceedings and Transactions of the British Entomological Society* 15: 55–67.

Stuble, K. L., Chick, L. D., Rodriguez-Cabal, M. A., Lessard, J.-P., and Sanders, N. J. 2013. Fire ants are drivers of biodiversity loss: a reply to King and Tschinkel (2013). *Ecological Entomology* 38: 540–542.

Sturgis, S. J., and Gordon, D. M. 2012 Nestmate recognition in ants (Hymenoptera: Formicidae): a review. *Myrmecological News* 16: 101–110.

Svanberg, I., and Berggren, A. 2019. Ant schnapps for health and pleasure: the use of *Formica rufa* L. (Hymenoptera: Formicidae) to flavour aquavit. *Journal of Ethnobiology* 15: doi: 10.1186/s13002-019-0347-7.

Swammerdam, J. 1758. *The Book of Nature; of the history of insects; reduced to distinct classes, confirmed by particular instances, displayed in the anatomical analysis of many species … etc.* London: Seyffert.

Sweeney, R. C. H. 1950. Identification of British ants (Hym.: Formicidae) with keys to genera and species. *Entomologist's Gazette* 1: 64–83.

Tartally, A., Nash, D. R., and Varga, Z. 2019. Changes in host ant communities of alcon blue butterflies in abandoned mountain hay meadows. *Insect Conservation and Diversity* 12: 492–500.

Thomas, J. A., and Emmet, A. M. 1989. *Maculinea arion.* In: *The Moths and Butterflies of Great Britain and Ireland.* Volume 7, Part 1, pp. 171–175. Colchester: Harley Books.

Thomas, J. A., and Lewington, R. 2010. *The Butterflies of Britain and Ireland.* London: British Wildlife Publishing.

Thomas, J. A., and Schönrogge, K. 2019. Conservation of co-evolved interactions: understanding the *Maculinea–Myrmica* complex. *Insect Conservation and Diversity* 12: 459–466.

Torres Minguez, A. 1924. Notas malacológicas. *Butlletí de la Institució Catalana d'Historia Natural* 24: 104–114.

Trible, W., and Kronauer, J. C. 2017. Caste development and evolution in ants: it's all about size. *Journal of Experimental Biology* 220: 53–62.

Tschinkel, W. R. 2004. The nest architecture of the Florida harvester ant, *Pogonomyrmex badius. Journal of Insect Science* 4: 21. doi: 10.1093/jis/4.1.21.

Tucker, C. M. 1889. *Fairy Frisket; or Peeps at Insect Life.* London: Nelson.

Tumlinson, J. H., Silverstein, R. M., Moser, J. C., Brownlee, R. G., and Ruth, J. M. 1971. Identification of the trail pheromone of a leaf-cutting ant, *Atta texana. Nature* 234: 348–349.

Van Mele, P. 2007. A historical review of research on the weaver ant *Oecophylla* in biological control. *Agricultural and Forest Entomology* 10: 13–22.

Virtanen, T., and Neuvonen, S. 1999. Climate change and macrolepidopteran biodiversity in Finland. *Chemosphere Global Change Science* 1: 439–448.

von Heyden, C. 1823. Ueber ein sonderbar gestaltetes Thierchen. *Oken's Isis* 2: 1247–1249.

von Spix, J. B. 1824. Protocolle der öffentlichen Sitzungen der math.-physik. *Bayerische Akademie der Wissenschaften Munchen* 11: 121–124.

Wagner, B., Tissot, M., Cuevas, W., and Gordon, D. M. 2000. Harvester ants utilize cuticular hydrocarbons in nestmate recognition. *Journal of Chemical Ecology* 26: 2245–2257.

Wallace, A. R. 1853. *Narrative of Travels on the Amazon and Rio Negro, with an account of the native tribes, and observations on the climate, geology, and natural history of the Amazon Valley.* London: Reeve.

Wasmann, E. 1894. Kritisches Verzeichniss der myrmecophilen und termitophilen. *Deutsche Entomologische Zeitschrift* 39: 41–44.

Watanabe, S., Murakami, T., Yoshimura, J., and Hasegawa, E. 2016. Color polymorphism in an aphid is maintained by attending ants. *Science Advances* 2: e1600606.

Watson, H. C. 1852. *Cybele Britannica or British Plants and Their Geographical Relations. Volume 3. Distribution of Species.* London: Longman.

Way, M. J., and Khoo, K. C. 1992. Role of ants in pest management. *Annual Review of Entomology* 37: 479–503.

Wegnez, P., Ignace, D., Fichefet, V., Hardy, M., Plume, T., and Timmermann, M. 2012. *Fourmis de Wallonie (2003–2011)*. Gembloux: Département de l'Étude du Milieu Naturel et Agricole.

Wells, H. G. 1895. A moth – genus novo. *Pall Mall Gazette*, 28 March.

Wells, H. G. 1898. *War of the Worlds*. London: William Heinemann.

Wells, H. G. 1903. Valley of the spiders. *Pearson's Magazine* pp. 298–304.

Wells, H. G. 1904a. *The Food of the Gods and How it Came to Earth*. London: Macmillan.

Wells, H. G. 1904b. Empire of the ants. *The Strand Magazine* pp. 1–9.

Westwood, J. O. 1839–1840. *An Introduction to the Modern Classification of Insects; founded on the natural habits and corresponding organisation of the different families*. 2 vols. London: Longman, Orme, Brown, Green, & Longmans.

Wetterer, J. K. 2010. Worldwide spread of the pharaoh ant, *Monomorium pharaonis* (Hymenoptera: Formicidae). *Myrmecological News* 13: 115–129.

Wetterer, J. K., Wild, A. L., Suarez, A. V., Roura-Pascual, N., and Espaldier, X. 2009. Worldwide spread of the Argentine ant, *Linepithema humile* (Hymenoptera: Formicidae). *Myrmecological News* 12: 187–194.

Wheeler, W. M. 1911. The ant-colony as an organism. *Journal of Morphology* 22: 307–325.

Wheeler, W. M. 1913. A solitary wasp (*Aphilanthops frigidus* F. Smith) that provisions its nest with queen ants. *Journal of Animal Behaviour* 3: 374–387.

White, T. H. 1958. *The Once and Future King*. London: Collins.

White, W. F. 1895. *Ants and Their Ways, with illustrations, and an appendix giving a complete list of genera and species of the British ants*. London: Religious Tract Society.

Williams, C. B. 1960. The range and pattern of insect abundance. *American Naturalist* 94: 137–151.

Wilson, E. O. 1987. The little things that run the world (the importance and conservation of invertebrates). *Conservation Biology* 1: 344–346.

Wilson, E. O., and Taylor, R. W. 1964. A fossil ant colony: new evidence of social antiquity. *Psyche* 71: 93–103.

Wilson, E. O., Carpenter, F. M., and Brown, W. L. Jr. 1967. The first Mesozoic ants. *Science* 157: 1038–1040.

Wood, J. G. 1903. *Nature's Teachings: Human Invention Anticipated by Nature*. London: Glaisher.

Wray, J. 1670. Extract of a letter, written by Mr. John Wray to the Publisher January 13, 1670. Concerning some un-common observations and experiments made with the acid juyce to be found in ants. *Philosophical Transactions* 5: 2063–2066.

Yarrow, I. H. H. 1953. A book on ants. *Ants*. By Derek Wragge Morley [book review]. *The Listener* 8 October 1953.

Yarrow, I. H. H. 1954. The British ants allied to *Formica fusca* L. (Hym., Formicidae). *Transactions of the Society for British Entomology* 11: 229–244.

Zeil, J., and Fleischmann, P. N. 2019. The learning walks of ants (Hymenoptera: Formicidae). *Myrmecological News* 29: 93–110.

Zoebelein, G. 1956. Der Honigtau als Nahrung der Insekten. II. *Zeitschrift für Angewandte Entomologie* 39: 129–167.

Zollikofer, C. P. E. 1994a. Stepping patterns in ants. I. Influence of speed and curvature. *Journal of Experimental Biology* 192: 95–106.

Zollikofer, C. P. E. 1994b. Stepping patterns in ants. II. Influence of body morphology. *Journal of Experimental Biology* 192: 107–118.

Zollikofer, C. P. E. 1994c. Stepping patterns in ants. III. Influence of load. *Journal of Experimental Biology* 192: 119–127.

Illustration credits

All photographs and figures are copyright © the author or are understood to be out of copyright, except for those listed below.

Bloomsbury Publishing would like to thank those listed below for providing illustrations and for permission to reproduce copyright material within this book. While every effort has been made to trace and acknowledge copyright holders, we would like to apologise for any errors or omissions, and invite readers to inform us so that corrections can be made in any future editions.

1 © Brian Valentine; 2–3 © Kim Taylor/Nature Picture Library; 6 © Andy Sands/ Nature Picture Library; 12 inset © Penny Metal; 13 © N Mrtgh/Shutterstock.com; 14 © Penny Metal; 16 © Ken Griffiths/Shutterstock.com; 18, 21, 22 © Alex Wild; 25 © Tim Bingham/iStock.com; 30 © Brian Valentine; 32 © Julian Baker (JB Illustrations); 33 © Penny Metal; 35 © Alex Wild; 38 top © Pedro Bernardo/ Shutterstock.com, bottom © Julian Baker (JB Illustrations); 39 © gstalker/ Shutterstock.com; 41 top left and bottom right © Brian Valentine, top right © Penny Metal, middle left © IanRedding/Shutterstock.com, middle right and bottom left © Tristan Bantock; 42 © Julian Baker (JB Illustrations); 46 © kingma photos/Shutterstock.com; 47 © Julian Baker (JB Illustrations); 49 © Brian Valentine; 51 © Alex Wild; 53 © Kim Taylor/Nature Picture Library; 55 © Alex Wild; 57 top left, top right and bottom right © Penny Metal, middle left © Nature Photographers Ltd/Alamy Stock Photo, middle right and bottom left © Tristan Bantock; 58 © Tapui/Shutterstock.com; 61 © Tristan Bantock; 62 © Paul Sterry/ Nature Photographers Ltd; 64 © Matt Hamer; 69 'Ponera coarctata-tandem run' © Michal Kukla/Antwiki (CC BY-SA 4.0); 71 © Heather Broccard-Bell/ iStock.com; 72 'Tapinoma erraticum worker with brood' © Phil Honle/AntWiki (CC BY-SA 4.0); 74 © Fabio Sacchi/Shutterstock.com; 75 © Stewart Taylor; 76 © Richard Becker/Alamy Stock Photo; 77 © Suman Ghosh/iStock.com; 78, 79 © Penny Metal; 82 © Ihor Hvozdetskyi/Shutterstock.com; 85 'The thief ant Solenopsis fugax stealing Tapinoma brood' © Phil Honle/AntWiki (CC BY-SA 4.0); 86 © Cherdchai Chaivimol/Shutterstock.com; 87 'Worker ant of Stenamma debile' © Arnstein Staverløkk/Norsk institutt for naturforskning (CC BY-SA 4.0); 88 'Strongylognathus testaceus queen' © Phil Honle/AntWiki (CC BY-SA 4.0); 89 © thatmacroguy/Shutterstock.com; 91 © Matt Hamer; 93 top © Andy Marquis, bottom 'Specimen: CASENT0901496 Cardiocondyla britteni - dorsal view' © Ryan Perry/AntWeb.org (CC BY 4.0); 96 © Tristan Bantock; 97 © Alex Hyde; 105 © thatmacroguy/Shutterstock.com; 106 © Pavel Krasensky/Shutterstock.com; 109 © Ant Cooper/Shutterstock.com; 110 © Pavel Krasensky/Shutterstock.com; 112 © Penny Metal; 115 © Bjoern Wylezich/Shutterstock.com; 116 left © The Natural History Museum/Alamy Stock Photo, right © Custom Life Science Images/Alamy Stock Photo; 121 © Penny Metal; 123 right 'Specimen: AMNH-

Acknowledgements

No book is written in a vacuum, even one finished in the back living room during the strange self-secluded lockdowns of 2020 and 2021. During this very weird time my family supported me endlessly. Catrina Ure helped me find a calm space, both physical and mental, in which to write, and kept me focused if I got too distracted by chickens in the garden or a cat stuck on the roof. Through this writing I have found a spiritual solace in a natural world that had recently become difficult to access in person. My daughters Lillian and Verity Ure-Jones have been my first editors, working their way through some complex science and some self-indulgent reminiscences to make sure I didn't go off-piste too often and that my thread remained followable even in the dark labyrinth of the ant nest. Lillian also helped with the translation of some interesting technical German. Verity drew draft line illustrations for the identification key, and several other pictures in the book. And my son Calvin Ure-Jones took me out on regular afternoon cycle rides through South London to get the breeze through my hair again, though we never encountered flying ant day.

Many people contributed ideas, thoughts, support, photographs and good wishes. In particular, Brian Valentine has given more than his fair share of superb ant photographs, mostly taken in his back garden, and they lift off the pages in wonderful clarity and depth to give the book a vibrancy I could not have hoped for when I started out.

My grateful thanks also go to: Gail Ashton, Rich Austin, Tristan Bantock, Tim Barnes, Roland Bates, Ian Beavis, Pete Brash, Thomas Clements, Matthew Cobb, Brian Eversham, Steven Falk, Mathew Frith, Chris Goodlad, Marcus Grenfeld, Matt Hamer, Naomi Harvey, Neil Hulme, Alex Hyde, Simon Leather, Peter Llewellyn, Andy Marquis, Scott Mayson, Erica McAlister, Penny Metal, Matthew Oates, Joe Parker, Mo Richards, Simon Robson, Andrew Salisbury, David Sargan, Chris Shortall, Sally-Ann Spence, Walter Tschinkel, Claire Waller, Alex Wild and Richard Wilson.

Finally, my thanks go to the editors, Katy Roper and Hugh Brazier, who painstakingly chiselled away many of the rough edges from my text and shone stark light into the hazy shadows of my image selections.

Index

Page numbers in **bold** refer to illustrations.
Page numbers in *italics* refer to tables.